MEI structured mathematics

Mechanics 3

JOHN BERRY
PAT BRYDEN
TED GRAHAM
ROGER PORKESS

Series Editor: Roger Porkess

MEI Structured Mathematics is supported by industry:
BNFL, Casio, GEC, Intercity, JCB, Lucas, The National Grid Company, Sharp, Texas Instruments, Thorn EMI

Hodder & Stoughton
A MEMBER OF THE HODDER HEADLINE GROUP

Acknowledgements

We are grateful to the following companies, individuals and institutions who have given permission to reproduce photographs in this book. Every effort has been made to trace and acknowledge ownership of copyright. The publishers will be glad to make suitable arrangements with any copyright holder whom it has not been possible to contact.

J. Allan Cash (2 both, 71 top left, 124 left); Ed Buziak (123); Mamas & Papas Ltd (58); Pekka Parviainen/Science Photo Library (71 bottom); Tanya Piejus (69 both); Helene Rogers/TRIP (71 right, 44); Robert Harding Picture Library (122); Royal Observatory, Edinburgh/Science Photo Library (121); Colin Taylor Productions (16, 60, 61, 112 centre, 124).

British Library Cataloguing in Publication Data

Mechanics. – Book 3. – (MEI Structured
Mathematics Series)
I. Berry, John II. Series
530

iSBN 0–340–57862–9

First published 1995
Impression number 10 9 8 7 6 5 4 3 2 1
Year 1998 1997 1996 1995

Typeset by Wearset, Boldon, Tyne and Wear.
Printed in Great Britain for Hodder & Stoughton Educational, a division of Hodder Headline Plc, 338 Euston Road, London NW1 3BH by Bath Press Ltd.

MEI Structured Mathematics

Mathematics is not only a beautiful and exciting subject in its own right but also one that underpins many other branches of learning. It is consequently fundamental to the success of a modern economy.

MEI Structured Mathematics is designed to increase substantially the number of people taking the subject post-GCSE, by making it accessible, interesting and relevant to a wide range of students.

It is a credit accumulation scheme based on 45 hour components which may be taken individually or aggregated to give:

3 components	AS Mathematics
6 components	A Level Mathematics
9 components	A Level Mathematics + AS Further Mathematics
12 components	A Level Mathematics + A Level Further Mathematics

Components may alternatively be combined to give other A or AS certifications (in Statistics, for example) or they may be used to obtain credit towards other types of qualification.

The course is examined by the Oxford and Cambridge Schools Examination Board, with examinations held in January and June each year.

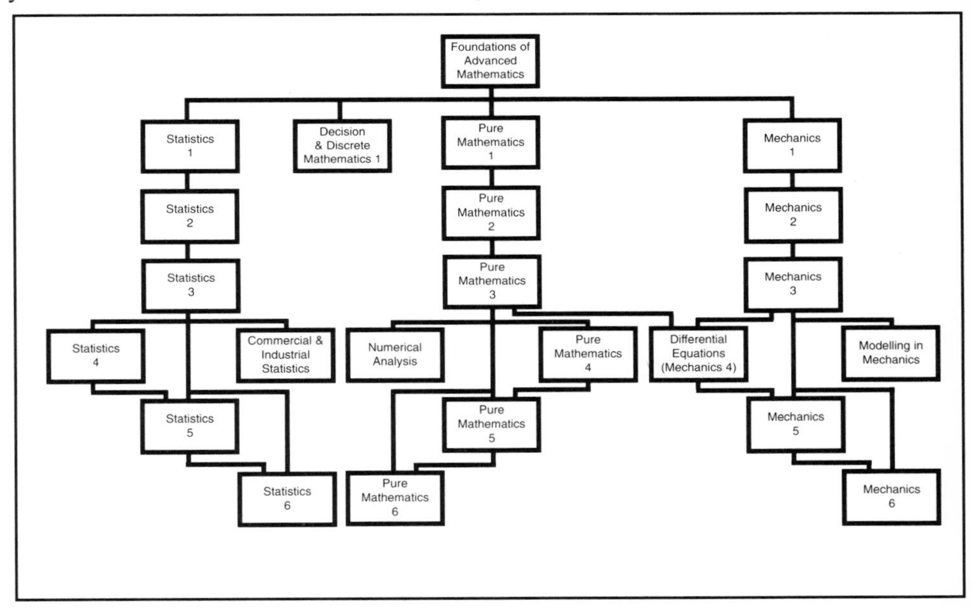

This is one of the series of books written to support the course. Its position within the whole scheme can be seen in the diagram above.

Mathematics in Education and Industry is a curriculum development body which aims to promote the links between Education and Industry in Mathematics and to produce relevant examination and teaching syllabuses and support material. Since its foundation in the 1960s, MEI has provided syllabuses for GCSE (or O Level), Additional Mathematics and A Level.

For more information about MEI Structured Mathematics or other syllabuses and materials, write to MEI Office, 11 Market Street, Bradford on Avon, BA15 ILL.

Introduction

This is the third in the series of books written to support the Mechanics components of MEI Structured Mathematics, but you may also use them for an independent course of study in the subject. For those who take an A Level consisting of Pure Mathematics and Mechanics only, this book completes the Mechanics part of the course.

In the past, Mechanics has often been presented as algebraic manipulation with little reference to the real world. The books in this series have been designed to present the subject as a powerful and exciting means of modelling real situations. The many applications of mechanics make it an important subject for applied mathematicians, scientists and engineers, but to use mechanics on real problems you must also understand the whole modelling procedure, which allows you to relate your knowledge of mathematics to the world around you. These books help you to do this.

In *Mechanics 3* you meet circular motion, elasticity and simple harmonic motion for the first time, and use calculus methods to extend earlier work on centre of mass. The book, and the basic Mechanics course, is completed by the final chapter on Dimensional Analysis.

We would like to thank the many people who have given help and advice with this book as it has developed, and also the various examination boards who have allowed us to use their questions in the exercises.

John Berry
Pat Bryden
Ted Graham
Roger Porkess

Contents

1 Circular motion

Whirlpools and storms his circling arm invest
With all the might of gravitation blest.

Alexander Pope

These pictures show some objects which move in circular paths. What other examples can you think of.

What makes objects move in circles?
Why does the moon circle the earth?
What happens to a "hammer" when an athlete lets it go?
Does the pilot of the aeroplane need to be strapped into his seat at the top of the loop in order not to fall out?

The answers to these questions lie in the nature of circular motion. Even if an object is moving at constant speed in a circle, its velocity keeps changing because its direction of motion keeps changing. Consequently the object is accelerating, and so according to Newton's Second Law there must be a force acting on it. The force required to keep an object moving in a circle can be provided in many ways.

Without the Earth's gravitational force, the Moon would move off at constant speed in a straight line into space. The wire attached to the athlete's hammer provides a tension force which keeps the ball moving in a circle. When the athlete lets go, the ball flies off at a tangent because the tension has disappeared.

No upward force is necessary to stop the pilot falling out of the plane because his weight contributes to the force required for motion in a circle.

In this chapter, these effects are explained.

Notation

To describe circular motion (or indeed any other topic) mathematically you need a suitable notation. It will be helpful in this chapter to use the notation (attributed to Newton) for differentiation with respect to time in which, for example, $\frac{\mathrm{d}s}{\mathrm{d}t}$ is written as $\dot{s}$, and $\frac{\mathrm{d}^2\theta}{\mathrm{d}t^2}$ as $\ddot{\theta}$.

Figure 1.1 shows a particle P moving round the circumference of a circle of radius r, centre O. At time t, the position vector $\overrightarrow{\mathrm{OP}}$ of the particle makes an angle θ (in radians) with the fixed direction $\overrightarrow{\mathrm{OA}}$. The arc length AP is denoted by s.

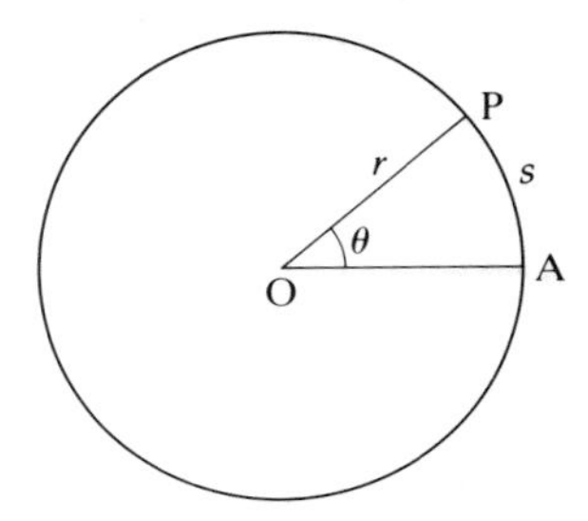

Figure 1.1

Angular speed

Using this notation,

$$s = r\theta$$

Differentiating this with respect to time using the product rule gives the rate at which this arc length is increasing:

$$\frac{\mathrm{d}s}{\mathrm{d}t} = r\frac{\mathrm{d}\theta}{\mathrm{d}t} + \theta\frac{\mathrm{d}r}{\mathrm{d}t}.$$

Since r is constant for a circle, $\frac{\mathrm{d}r}{\mathrm{d}t} = 0$, and

$$\frac{\mathrm{d}s}{\mathrm{d}t} = r\frac{\mathrm{d}\theta}{\mathrm{d}t} \text{ or } \dot{s} = r\dot{\theta}. \qquad (1)$$

In this equation $\dot{s}$ is the speed at which P is moving round the circle (often denoted by v), and $\dot{\theta}$ is the rate at which the angle θ is increasing, i.e. the rate at which the position vector $\overrightarrow{\mathrm{OP}}$ is rotating.

The quantity $\frac{\mathrm{d}\theta}{\mathrm{d}t}$, or $\dot{\theta}$, can be called the *angular velocity* or the *angular speed* of P. In more advanced work, angular velocity is treated as a vector, whose direction is taken to be that of the axis of rotation. In this book, $\dfrac{d\theta}{dt}$ is treated as a scalar and referred to as angular speed, but is given a sign: positive for an anticlockwise rotation and negative for a clockwise rotation.

Angular speed is often denoted by ω, the Greek letter omega. So the equation $\dot{s} = r\dot{\theta}$ may be written as

$$v = r\omega$$

Notice that for this equation to hold, θ must be measured in radians, so the angular speed is measured in *radians per second* or rad s^{-1}.

NOTE

It is customary to quote angular speeds as multiples of π unless otherwise requested.

Figure 1.2 shows a disc rotating about its centre, O, with angular speed ω. The line OB represents any radius.

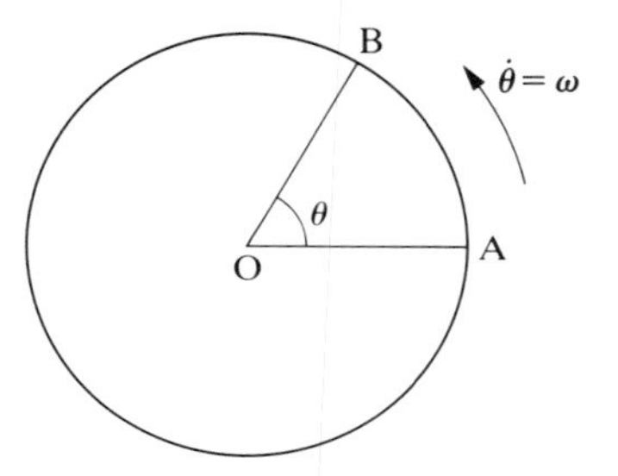

Figure 1.2

Every point on the disc describes a circular path, and all points have the same angular speed. However the *actual* speed of any point depends on its distance from the centre: increasing r in the equation $v = r\omega$ increases v. You will appreciate this if you have ever been at the end of a rotating line of people in a dance or watched a body of marching soldiers wheeling round a corner.

Angular speeds are sometimes measured in revolutions per second or revolutions per minute (rpm) where one revolution is equal to 2π radians. For example, turntables for records rotate at 45 or $33\frac{1}{3}$ rpm, and at cruising speeds crankshafts in car engines typically rotate at 3000 to 4000 rpm.

EXAMPLE

A police car drives at 40 mph around a roundabout in a circle of radius 16 m. A second car moves so that it has the same angular speed as the police car but in a circle of radius 12 m. Is the second car breaking the 30 mph speed limit? (Use the approximation 1 mile $= \frac{8}{5}$ km.)

Solution

Converting miles per hour to metres per second gives:

$$40\,\text{mph} = 40 \times \tfrac{8}{5}\text{km per hour}$$

$$= \frac{40 \times 8 \times 1000}{5 \times 3600}\,\text{ms}^{-1}$$

$$= \frac{160}{9}\,\text{ms}^{-1}$$

Using $v = r\omega$,

$$\omega = \frac{160}{9 \times 16}\,\text{rads}^{-1}$$

$$= \frac{10}{9}\,\text{rads}^{-1}$$

The speed of the second car is:

$$v = 12\omega$$

$$= \frac{10}{9} \times 12\,\text{ms}^{-1}$$

$$= \frac{120 \times 5 \times 3600}{9 \times 8 \times 1000}\,\text{mph}$$

$$= 30\,\text{mph}$$

The second car is just on the speed limit.

NOTE

A quicker way to do this question would be to notice that, because the cars have the same angular speed, the actual speeds of the cars are proportional to the radii of the circles in which they are moving. Using this method it is possible to stay in mph. The ratio of the two radii is $\frac{12}{16}$ so the speed of the second car is $\frac{12}{16} \times 40\,\text{mph} = 30\,\text{mph}$.

Exercise 1A

1. Find the angular speed, in radians per second correct to one decimal place, of records rotating at
   (i) 78 rpm (ii) 45 rpm (iii) 33 rpm.
2. A big wheel at a fairground completes one revolution in 3 seconds. Find its angular speed (assumed constant), correct to 1 decimal place, in (i) rpm (ii) radians per second.
3. A flywheel is rotating at 300 radians per second. Express this angular speed in rpm, correct to the nearest whole number.
4. A record on a turntable is rotating at 33 rpm. What is the speed in ms^{-1} of a point on the circumference of the record if its diameter is
   (i) 12 inches (ii) 10 inches?

Exercise 1A continued

5. A lawnmower engine is started by pulling a rope that has been wound round a cylinder of radius 4 cm. Find the angular speed of the cylinder at a moment when the rope is being pulled with a speed of $1.3\,\text{ms}^{-1}$. Give your answers in radians per second, correct to one decimal place.

6. The wheels of a car have radius 20 cm. What is the angular speed, in radians per second correct to one decimal place, of a wheel when the car is travelling at
(i) $10\,\text{ms}^{-1}$ (ii) $30\,\text{ms}^{-1}$?

7. What is the average angular speed of the Earth in radians per second as it: (i) orbits the sun; (ii) rotates about its own axis?

The radius of the Earth is 6400 km.
(iii) At what speed is someone on the equator travelling relative to the centre of the Earth?
(iv) At what speed are you travelling relative to the centre of the Earth?

8. A tractor has front wheels of diameter 70 cm and back wheels of diameter 1.6 m. What is the ratio of their angular speeds when the tractor is being driven along a straight road?

9. (i) Find the kinetic energy of a 50 kg person riding a big wheel with radius 5 m when the ride is rotating at 3 rpm. You should assume that the person can be modelled as a particle.
(ii) Explain why this modelling assumption is necessary.

10. The minute hand of a clock is 1.2 m long and the hour hand is 0.8 m long.
(i) Find the speeds of the tips of the hands.
(ii) Find the ratio of the speeds of the tips of the hands and explain why this is not the same as the ratio of the angular speeds of the hands.

11. The diagram represents a "Chairoplane" ride at a fair: it completes one revolution every 2.5 seconds.
(i) Find the radius of the circular path which a rider follows.
(ii) Find the speed of a rider.

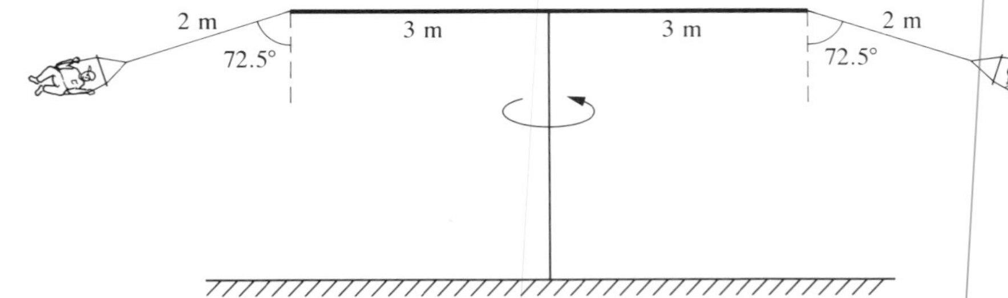

12. The position vector of a rider on a helter-skelter is given by

$$\mathbf{r} = 2\sin t\,\mathbf{i} + 2\cos t\,\mathbf{j} + (8 - \tfrac{1}{2}t)\mathbf{k}$$

where the units are in metres and seconds. The unit vector **k** is vertically upwards.

(i) Find an expression for the velocity of the rider at time t.

(ii) Find the speed of the rider in terms of t.

(iii) Find the magnitude and direction of the rider's acceleration when $t = \frac{\pi}{4}$.

13. The tape in a cassette player passes over the heads at a speed of $5\,\text{cm s}^{-1}$ and is taken up on a spool of diameter $1.7\,\text{cm}$. The thickness of the tape is $0.01\,\text{mm}$. The take-up spool is initially empty. Find an expression for the angular speed of the take-up spool at time t seconds. What modelling assumptions have you made?

14. The diagram shows a roundabout in a playground, seen from above. It is rotating clockwise. A child on the roundabout, at X, aims a ball at a friend sitting opposite at Y.

(i) Once the ball is thrown, can the friend catch it?

(ii) Draw a plan of the path of the ball after it has been thrown.

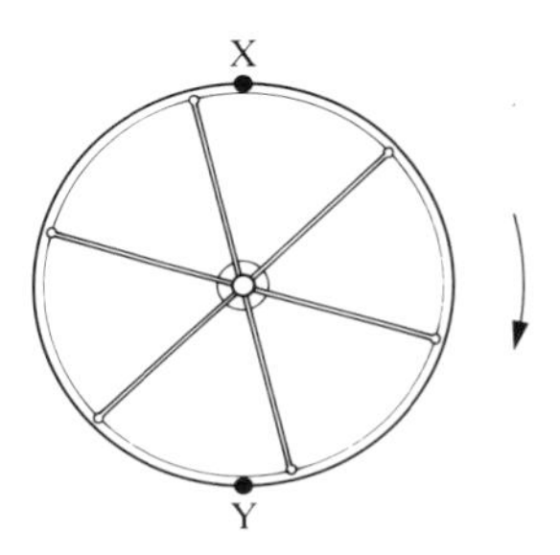

Velocity and acceleration

Velocity and acceleration are both vector quantities. They can be expressed either in magnitude-direction (polar) form, or in components. When describing circular motion or other orbits it is most convenient to take components in directions along the radius (*radial* direction) and at right angles to it (*transverse* direction).

For a particle moving round a circle of radius r, the velocity has:

radial component: 0
transverse component: $r\dot{\theta}$ or $r\omega$.

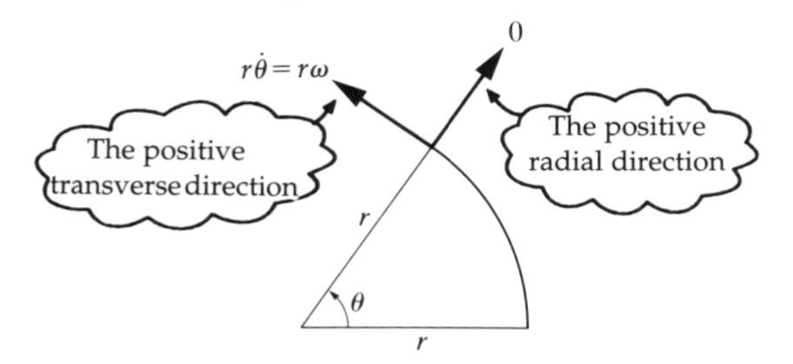

Figure 1.3 Velocity

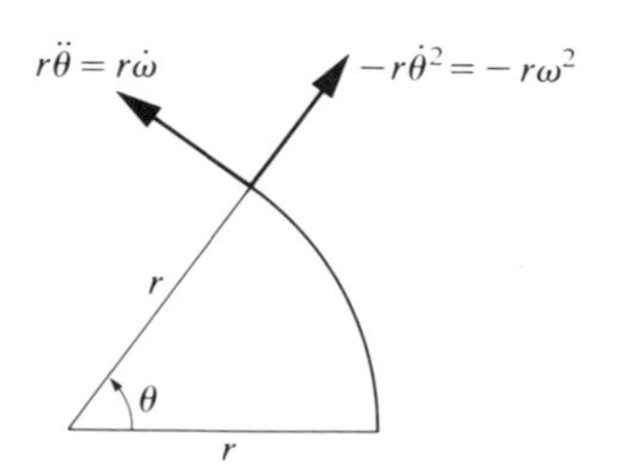

Figure 1.4 Acceleration

The acceleration of a particle moving round a circle of radius r is found to have:

radial component: $-r\dot{\theta}^2$ or $-r\omega^2$
transverse component: $r\ddot{\theta}$ or $r\dot{\omega}$.

The transverse component is just what you would expect, the radius multiplied by the angular acceleration, $\ddot{\theta}$. If the particle has constant angular speed, its angular acceleration is zero and so the transverse component of its acceleration is also zero.

By contrast, the radial component of the acceleration, $-r\omega^2$, is almost certainly not a result you would have expected intuitively. It tells you that a particle travelling in a circle is always accelerating towards the centre of the circle, but without ever getting any closer to the centre. If this seems a strange idea, you may find it helpful to remember that circular motion is not a natural state; left to itself a particle will travel in a straight line. To keep a particle in the unnatural state of circular motion it must be given an acceleration at right angles to its motion, i.e. towards the centre of the circle.

The derivation of these expressions for the acceleration of a particle in circular motion is complicated by the fact that the radial and transverse directions are themselves changing as the particle moves round the circle, in contrast to the fixed x and y directions in the Cartesian system. The derivation is given on page 158 in the section entitled Mathematical Notes. At first reading you may prefer to accept the results, but make sure that at a later stage you work through and understand the derivation.

Circular motion with constant speed

In this section, the circular motion is assumed to be uniform and so have no transverse component of acceleration. Later in the chapter (page 25 onwards), situations are considered in which the angular speed varies.

Problems involving circular motion often refer to the actual speed of the object, rather than its angular speed. It is easy to convert the one into the other using the relationship $v = r\omega$. This relationship can also be used to express the magnitude of the acceleration in terms of v and r.

$$\omega = \frac{v}{r}$$

$$a = r\omega^2 = r\left(\frac{v}{r}\right)^2$$

$$\Rightarrow \quad a = \frac{v^2}{r} \text{ towards the centre.}$$

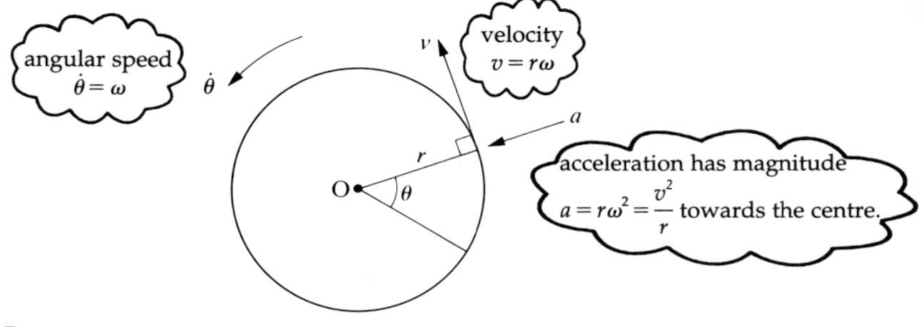

Figure 1.5

EXAMPLE

A fly is standing on a record at a distance of 8 cm from the centre. If the record is rotating at 45 rpm, find:

(i) the angular speed of the fly in radians per second;
(ii) the speed of the fly in metres per second;
(iii) the acceleration of the fly.

Solution

(i) One revolution is 2π radians so

$$45\,\text{rpm} = 45 \times 2\pi \text{ radians per minute}$$

$$= \frac{45 \times 2\pi}{60} \text{ radians per second}$$

$$= \frac{3\pi}{2} \text{ rad s}^{-1}.$$

(ii) If the speed of the fly is v ms^{-1}, v can be found using

$$v = r\omega$$

$$= 0.08 \times \frac{3\pi}{2}$$

$$= 0.377\ldots$$

So the speed of the fly is 0.38 ms^{-1} (to 2 decimal places).

(iii) The acceleration of the fly is given by

$$r\omega^2 = 0.08 \times \left(\frac{3\pi}{2}\right)^2 \text{ ms}^{-2}$$

$$= 1.78 \text{ ms}^{-2}.$$

It is directed towards the centre of the record.

The forces required for circular motion

Newton's First Law of motion states that a body will continue in a state of rest or uniform motion in a straight line unless acted upon by an external force. Any object moving in a circle, such as the police car and the fly in the examples on pages 4 and 9 must therefore be acted upon by a resultant force in order to produce the required acceleration towards the centre.

A force towards the centre is called a *centripetal* (centre-seeking) force. A resultant centripetal force is necessary for a particle to move in a circular path.

Experiment

1. The aim of this experiment is to study the motion of a coin on a rotating disc.

   Set up the apparatus as shown in the photograph.

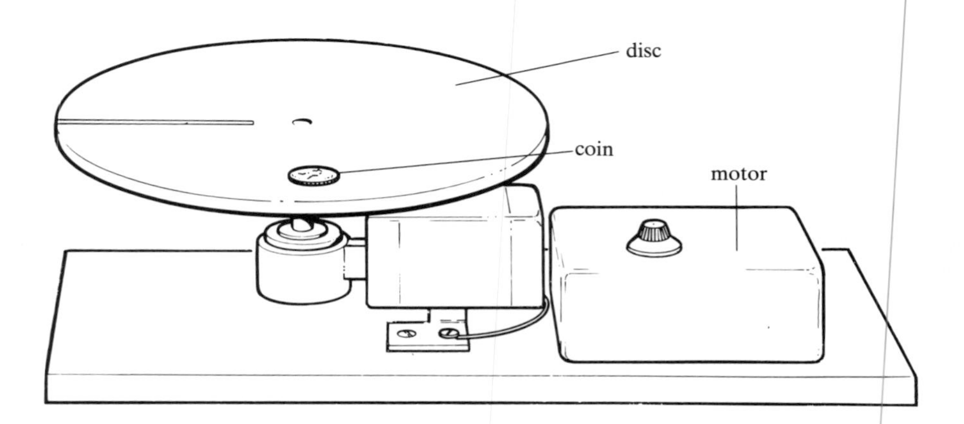

   Before you turn the motor on, answer the following questions about what you expect to happen.

   When the disc is rotating slowly
   (i) what will be the path of the coin?
   (ii) what forces will be acting on the coin?

   When the speed is increased.
   (iii) will the path of the coin still be the same?

   Are your answers affected by:
   (iv) the mass of the coin?
   (v) its distance from the centre of the turntable?

   Now turn the motor on and see what happens.

Here are two further experiments which may help you to interpret what is going on.

Experiment continued

2. Position four coins at equal intervals along a radius of the turntable. Turn on the motor so that the disc rotates but the coins remain at rest relative to the disc. Now gradually increase the motor speed. In what order do the coins slip?
3. Stick two coins together. Position a single coin and the two coins stuck together at the same distance from the centre of the turntable. Gradually increase the motor speed. Do they slip at the same time? Repeat the experiment a few times to check the consistency of your results. You could also try sticking three or four coins together.

Everyday examples of circular motion

You are now in a position to use Newton's Second Law to determine theoretical answers to some of the questions which were posed at the beginning of this chapter and within the investigation you have just done. These will, as usual, be obtained using models of the true motion. These models will be based on simplifying assumptions, for example zero air resistance.

The turntable

EXAMPLE

A coin is placed on a rotating turntable, 5 cm from the centre of rotation. The coefficient of friction between the coin and the turntable is 0.5.

(i) If the speed of rotation of the turntable is gradually increased, at what angular speed will the coin begin to slide?

(ii) What happens next?

Solution

(i) Because the speed of the turntable is increased only gradually, the coin will not slip tangentially.

The diagram shows the forces acting on the coin, and its acceleration.

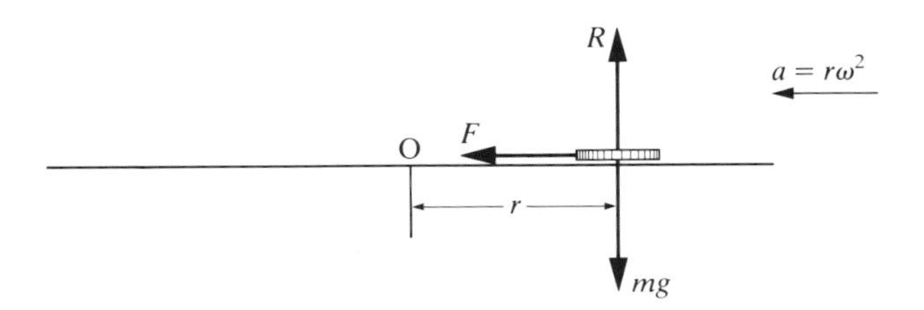

The acceleration is towards the centre of the circular path, O, so there must be a frictional force F in that direction.

There is no vertical component of acceleration, so the resultant force acting on the coin has no vertical component.

Therefore $$R - mg = 0$$
$$R = mg \qquad ①$$

By Newton's Second Law towards the centre of the circle:

$$F = ma = mr\omega^2 \qquad ②$$

The coin will not slide so long as $F \leq \mu R$.

Substituting from ② and ① this gives

$$mr\omega^2 \leq \mu m$$
$$\Rightarrow \quad r\omega^2 \leq \mu g$$

Notice that the mass m has been eliminated at this stage, so the answer does not depend upon it.

Taking g as 9.8 ms^{-2} and substituting for r and μ:

$$\omega^2 \leq 98$$
$$\omega \leq \sqrt{98}$$
$$\omega \leq 9.89\ldots$$

The coin will move in a circle provided that the angular speed is less than about 10 rad s^{-1} and this speed is independent of the mass of the coin.

(ii) When the angular speed increases beyond this, the coin slips to a new position. If the angular speed continues to increase it will slip right off the turntable. When it reaches the edge it will fly off in the direction of the tangent.

The conical pendulum

A conical pendulum consists of a small bob tied to one end of a string. The other end of the string is fixed and the bob is made to rotate in a horizontal circle below the fixed point so that the string describes a cone as in figure 1.6.

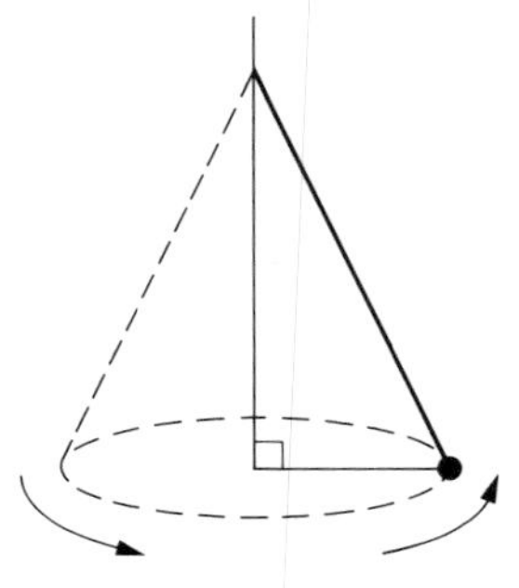

Figure 1.6

Experiment

Set up the conical pendulum apparatus shown in the illustration. If you do not have the equipment available you can perform this experiment using a bob suspended from your hand. You will probably have to move your hand in order to overcome the inevitable resistances and keep the bob rotating in the same circle, so this simpler apparatus is not absolutely equivalent to that in the illustration.

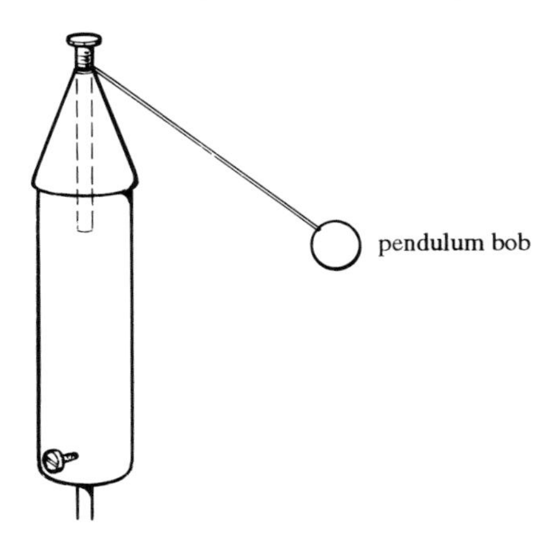

Before carrying out the experiment, write down the magnitude and direction of the acceleration of the bob, and draw a diagram showing the forces acting on the bob.

Now predict the answers to these questions.

1. How does the angle that the string makes with the vertical change when
   (i) the length of the string is increased?
   (ii) the angular speed is increased?
   (iii) the mass of the bob is increased?

2. Can the bob be made to rotate with the string at an angle of 90° or more?

Start up the motor and observe the motion of the bob once it has settled down, then carry out experiments to test your predictions.

Theoretical model for the conical pendulum

A conical pendulum may be modelled as a particle of mass m attached to a light, inextensible string of length l. The mass is rotating in a horizontal circle with angular speed ω, and the string makes an angle α with the downward vertical. The radius of the circle is r and the tension in the string is T, all in consistent units (e.g. SI units). The situation is shown in figure 1.7.

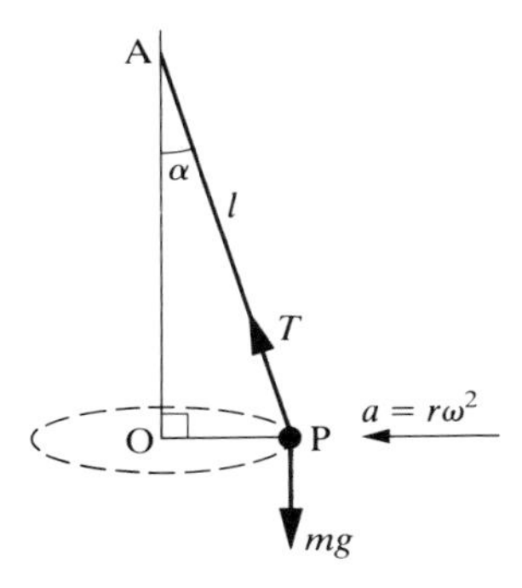

Figure 1.7

The magnitude of the acceleration is $r\omega^2$. The acceleration acts in a horizontal direction towards the centre of the circle. This means that there must be a resultant force acting towards the centre of the circle.

There are two forces acting on this particle, its weight mg and the tension T in the string.

As the acceleration of the particle has no vertical component, the resultant force has no vertical component, so

$$T\cos\alpha - mg = 0 \quad ①$$

Using Newton's Second Law towards the centre O of the circle:

$$T\sin\alpha = ma = mr\omega^2 \quad ②$$

In triangle AOP, $r = l\sin\alpha$

Substituting for r in ② gives $T\sin\alpha = m(l\sin\alpha)\omega^2$

$$\Rightarrow \quad T = ml\omega^2$$

Substituting this in ① gives

$$ml\omega^2\cos\alpha - mg = 0$$

$$\Rightarrow \quad \cos\alpha = \frac{g}{l\omega^2} \quad ③$$

This equation provides sufficient information to give theoretical answers to the questions in the experiment.

- When ω is kept constant and the length of the string is increased, equation ③ indicates that the value of $\cos\alpha$ decreases. But a decrease in $\cos\alpha$ entails an increase in α. So an increase in the length of the string will lead to an increase in α.
- When the length of the pendulum is unchanged, but the angular speed is increased, $\cos\alpha$ again decreases, leading to an increase in the angle α.
- The mass of the particle does not appear in equation ③, so it has no effect on the angle α.
- If $\alpha > 90°$, $\cos\alpha < 0$, so $\frac{g}{l\omega^2} < 0$, which is impossible. You can see from figure 1.7 that the tension in the string must have a vertical component to balance the weight of the particle.

EXAMPLE

The diagram, right, represents one of several arms of a fairground ride. The arm rotates about an axis and a rider is linked to the arm by a chain.

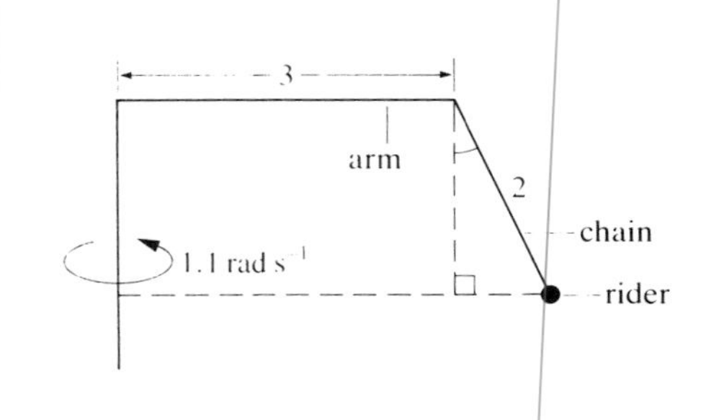

The chains are 2 m long and the arms are 3 m long. Find the angle that the chains make with the vertical when the rider rotates at $1.1\,\text{rad}\,\text{s}^{-1}$.

Solution

Let T be the tension in the chains holding a chair, and m kg the mass of chair and rider.

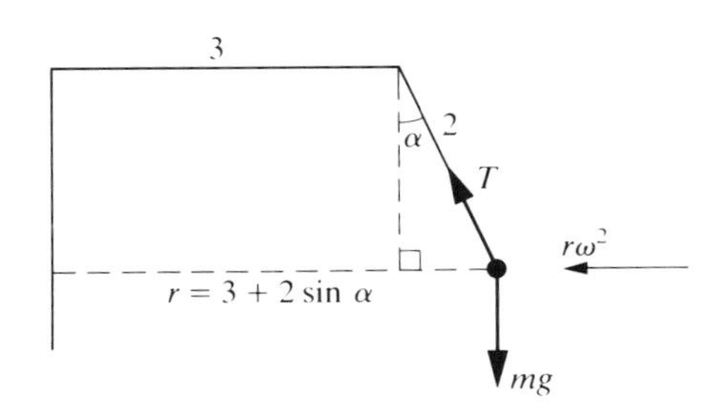

If the chains make an angle α with the vertical, the motion is in a horizontal circle of radius given by $r = 3 + 2\sin\alpha$. The magnitude of the acceleration is given by

$$r\omega^2 = (3 + 2\sin\alpha)\,1.1^2.$$

It is in a horizontal direction towards the centre of the circle. Using Newton's Second Law in this direction gives

$$F = ma = mr\omega^2$$

$$\Rightarrow \quad T\sin\alpha = m\,(3 + 2\sin\alpha)\,1.1^2$$

$$= 1.21m(3 + 2\sin\alpha) \qquad ①$$

Vertically $\qquad T\cos\alpha - mg = 0$

$$\Rightarrow \quad T = \frac{mg}{\cos\alpha}$$

Substituting for T in equation ①:

$$\frac{mg}{\cos\alpha}\sin\alpha = 1.21m\,(3 + 2\sin\alpha)$$

$$\Rightarrow \quad 9.8\tan\alpha = 3.63 + 2.42\sin\alpha$$

Since m cancels out at this stage, the angle does not depend on the mass of the rider

This equation cannot be solved directly, but a numerical method will give you the solution 25.5° correct to 3 significant figures. You might like to solve the equation yourself or check that this solution does in fact satisfy the equation.

NOTE

Since the answer does not depend on the mass of the rider and chair, when riders of different masses or even no riders, are on the equipment all the chains should make the same angle with the vertical.

Banked tracks

Activity

WARNING: keep away from other people and breakable objects when carrying out this activity.

Place a coin on a piece of stiff A4 card and hold it horizontally at arm's length with the coin near your hand.

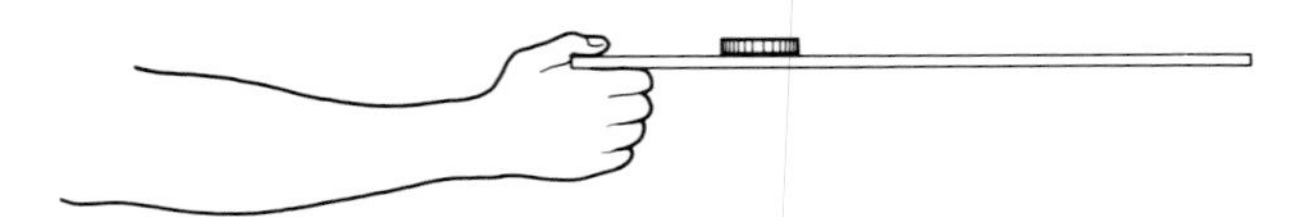

Turn round slowly so that your hand moves in a horizontal circle. Now gradually speed up. The outcome will probably not surprise you.

What happens, though, if you tilt the card?

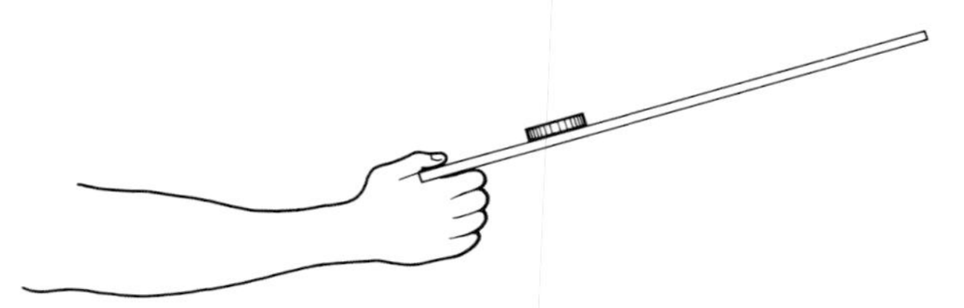

You may have noticed that when they curve round bends, most roads are banked so that the edge at the outside of the bend is slightly higher than that at the inside. For the same reason the outer rail of a railway track is slightly higher than the inner rail when it goes round a bend. On bobsleigh tracks the bends are almost bowl shaped, with a much larger gradient on the outside.

Figure 1.8 shows a car rounding a bend on a road which is banked so that the cross-section makes an angle α with the horizontal.

Figure 1.8

In modelling such situations, it is usual to treat the bend as part of a horizontal circle whose radius is large compared to the width of the car. In this case, the radius of the circle is taken to be r metres, and the speed of the car constant at v metres per second. The car is modelled as a particle which has an acceleration of $\frac{v^2}{r}$ ms^{-2} in a horizontal direction towards the centre of the circle. The forces and acceleration are shown in figure 1.9.

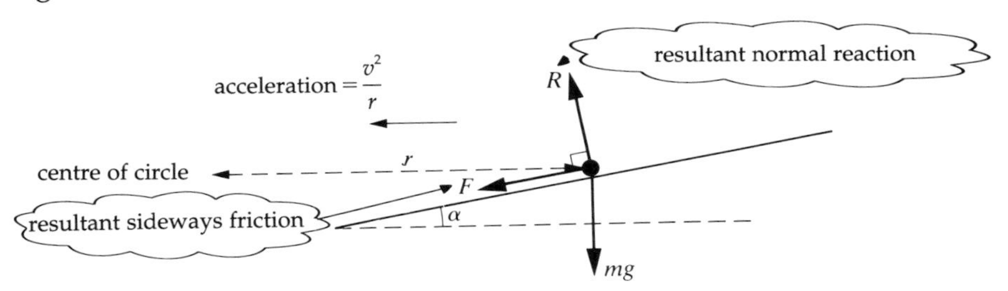

Figure 1.9

The direction of the frictional force F will be up or down the slope depending on whether the car has a tendency to slip sideways towards the inside or outside of the bend. Under what conditions do you think each of these will occur?

EXAMPLE

A car is rounding a bend of radius 100 m which is banked at an angle of 10° to the horizontal. At what speed must the car travel to ensure it has no tendency to slip sideways?

Solution

When there is no tendency to slip there is no frictional force, so in the plane perpendicular to the direction of motion of the car, the forces and acceleration are as shown. The only horizontal force is provided by the horizontal component of the normal reaction of the road on the car.

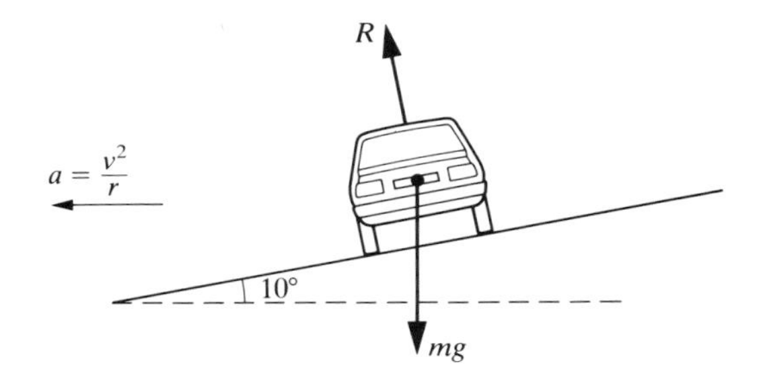

Vertically, there is no acceleration so there is no resultant force:

$$R\cos 10° - mg = 0$$

$$\Rightarrow \quad R = \frac{mg}{\cos 10°} \qquad \text{①}$$

By Newton's Second Law in the horizontal direction towards the centre of the circle:

$$R\sin 10° = ma = \frac{mv^2}{r}$$

$$= \frac{mv^2}{100}$$

Substituting for R from ①

$$\left(\frac{mg}{\cos 10°}\right)\sin 10° = \frac{mv^2}{100}$$

$$\Rightarrow \quad v^2 = 100g\tan 10°$$

$$\Rightarrow \quad v = 13.15\ldots$$

The mass, m, cancels out at this stage, so the answer does not depend on it.

The speed of the car must be about 13.15 ms^{-1}, or 30 mph.

There are two important points to notice in this example.

- The speed is the same whatever the mass of the car.
- The example was looking at the situation when the car does not tend to slide, and finding the speed at which this is the case. At this speed the car does not depend on friction to keep it from sliding, and indeed it could travel safely round the bend at this speed even in very icy conditions. However, at other speeds there is a tendency to slide, and friction actually helps the car to follow its intended path.

Safe speeds on a bend

What would happen in the previous example if the car travelled either more slowly than 13.15 ms^{-1} or more quickly?

The answer is that there would be a frictional force acting so as to prevent the car from sliding across the road.

There are two possible directions for the frictional force. When the vehicle is stationary or travelling slowly, there is a tendency to slide down the slope and the friction acts up the slope to prevent this. When it is travelling quickly round the bend, the car is more likely to slide up the slope, so the friction acts down the slope.

Fortunately, under most road conditions, the coefficient of friction between tyres and the road is large, typically about 0.8. This means that there is a range of speeds that are safe for negotiating any particular bend.

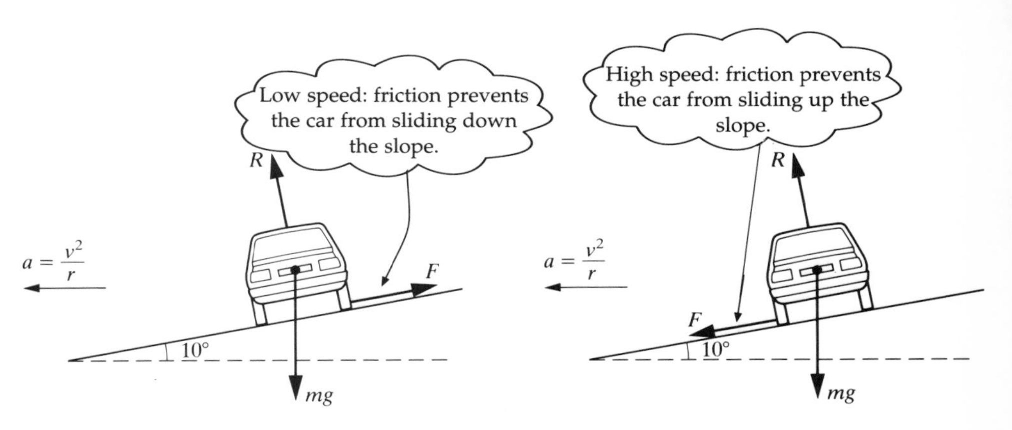

Investigation

In this investigation you should model the car as a particle.

(i) Show that a car will not slide up or down the slope provided

$$\sqrt{rg\frac{(\sin\alpha - \mu\cos\alpha)}{(\cos\alpha + \mu\sin\alpha)}} < v < \sqrt{rg\frac{(\sin\alpha + \mu\cos\alpha)}{(\cos\alpha - \mu\sin\alpha)}}$$

(ii) If $r = 100$ and $\alpha = 10°$ (so that $\tan\alpha = 0.176$) the minimum and maximum safe speeds in mph for different values of μ are given in the following table.

μ	0	0.1	0.2	0.3	0.4	0.5	0.6	0.7	0.8	0.9	1	1.1	1.2
Minimum safe speed	30	19	0	0	0	0	0	0	0	0	0	0	0
Maximum safe speed	30	37	44	50	55	61	66	70	75	80	84	89	93

Would you regard this bend as safe? How, by changing the values of r and α, could you make it safer?

EXAMPLE

A bend on a railway track has a radius of 500 m and is to be banked so that a train can negotiate the bend at 60 mph without the need for a lateral force between its wheels and the rail. The distance between the rails is 1.43 m.

How much higher should the outside rail be than the inside one?

Solution

There is very little friction between the track and wheels of a train. Any sideways force required is provided by the "lateral thrust" between the wheels and the rail. The ideal speed for the bend is such that the lateral thrust is zero.

The diagram shows the forces acting on the train and its acceleration when the track is banked at an angle α to the horizontal.

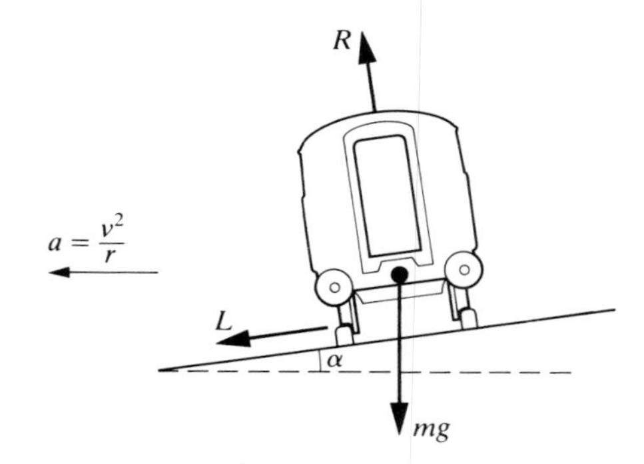

When there is no lateral thrust, $L = 0$.

Horizontally: $R\sin\alpha = \dfrac{mv^2}{r}$ ①

Vertically: $R\cos\alpha = mg$ ②

Dividing ① by ② gives

$$\tan\alpha = \frac{v^2}{rg}$$

Using the fact that $60\,\text{mph} = 26.8\,\text{ms}^{-1}$ this becomes

$$\tan\alpha = 0.147$$

$$\Rightarrow \quad \alpha = 8.4^\circ \text{ (to 2 significant figures)}$$

The outside rail should be raised by $1.43\sin\alpha$ metres, i.e. by about 21 cm.

Exercise 1B

1. The diagram shows two cars A and B travelling in different lanes (radii 24 m and 20 m) with speeds $18\,\text{ms}^{-1}$ (A) and $15\,\text{ms}^{-1}$ (B) round a roundabout.

Answer the following questions, giving reasons for your answers.
(i) Which car has the greater angular speed?
(ii) Is one car overtaking the other?
(iii) Find the magnitude of the acceleration of each car
(iv) In which direction is the resultant force on each car acting?

2. Two coins are placed on a horizontal turntable. Coin A has mass $15\,\mathrm{g}$ and is placed $5\,\mathrm{cm}$ from the centre; coin B has mass $10\,\mathrm{g}$ and is placed $7.5\,\mathrm{cm}$ from the centre. The coefficient of friction between each coin and the turntable is 0.4.
(i) Describe what happens to the coins when the turntable turns at
a) $6\,\mathrm{rad\,s^{-1}}$ b) $8\,\mathrm{rad\,s^{-1}}$ c) $10\,\mathrm{rad\,s^{-1}}$
(ii) What would happen if the coins were interchanged?

3. A car is travelling at a steady speed of $15\,\mathrm{ms^{-1}}$ round a roundabout of radius $20\,\mathrm{m}$.
(i) Criticise this false argument:
The car is travelling at a steady speed and so its speed is neither increasing nor decreasing and therefore the car has no acceleration.
(ii) Calculate the magnitude of the acceleration of the car.
(iii) The car has mass $800\,\mathrm{kg}$. Calculate the sideways force on each wheel assuming it to be the same for all four wheels.
(iv) Is the assumption in part (iii) realistic?

4. A fairground ride has seats at $3\,\mathrm{m}$ and at $4.5\,\mathrm{m}$ from the centre of rotation. Each rider travels in a horizontal circle. Say whether each of the following statements is true, giving your reasons:
(i) Riders in the two positions have the same angular speed at any time.
(ii) Riders in the two positions have the same speed at any time.
(iii) Riders in the two positions have the same magnitude of acceleration at any time.

5. A skater of mass $60\,\mathrm{kg}$ follows a circular path of radius $4\,\mathrm{m}$, moving at $2\,\mathrm{ms^{-1}}$.
(i) Calculate:
(a) the angular speed of the skater;
(b) the magnitude of the acceleration of the skater;
(c) the resultant force acting on the skater.
(ii) What modelling assumptions have you made?

6. Two spin driers, both of which rotate about a vertical axis, have different specifications as given in the table below.

Model	Rate of rotation	Drum diameter
A	600 rpm	60 cm
B	800 rpm	40 cm

State, with reasons, which model you would expect to be the more effective.

Exercise 1B continued

7. A satellite of mass M_s is in a circular orbit around the Earth with a radius of r metres. The force of attraction between the Earth and the satellite is given by

$$F = \frac{GM_E M_s}{r^2}$$

where $G = 6.67 \times 10^{-11}$ in SI units. The mass of the Earth M_E is 5.97×10^{24} kg.

(i) Find, in terms of r, expressions for
(a) its speed $v\,\text{ms}^{-1}$; (b) the time T it takes to complete one revolution.

(ii) Hence show that, for all satellites, T^2 is proportional to r^3.

A geostationary satellite orbits the Earth so that it is always above the same place on the equator.

(iii) How far is it from the centre of the Earth?

(The law found in part (ii) was discovered experimentally by Johannes Kepler (1571–1630) to hold true for the planets as they orbit the sun, and is commonly known as Kepler's third law.)

8. In this question you should assume that the orbit of the Earth around the sun is circular, with radius 1.44×10^{11} metres, and that the sun is fixed.

(i) Find the magnitude of the acceleration of the Earth as it orbits the sun.

The force of attraction between the Earth and the sun is given by

$$F = G\frac{M_E M_S}{r^2}$$

where M_E is the mass of the Earth, M_S is the mass of the sun, r the radius of the Earth's orbit and G the universal constant of gravitation.

(ii) Calculate the mass of the sun.

(iii) Comment on the significance of the fact that you cannot calculate the mass of the Earth from the radius of its orbit.

9. Sarah ties a model aeroplane of mass $180\,\text{g}$ to the end of a piece of string $80\,\text{cm}$ long and then swings it round so that the aeroplane travels in a horizontal circle. The aeroplane is not designed to fly and there is no lift force acting on its wings.

(i) Explain why it is not possible for the string to be horizontal.

Sarah gives the aeroplane an angular speed of $120\,\text{rpm}$.

(ii) What is the angular speed in radians per second?

(iii) Copy this diagram and mark in the tension, T, in the string, the weight of the toy and the direction of the acceleration.

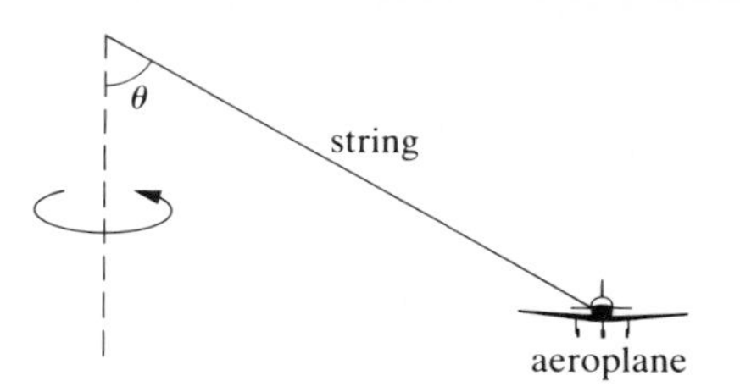

(iv) Write down the horizontal radial equation of motion for the toy and the vertical equilibrium equation in terms of the angle θ.
(v) Show that under these conditions θ has a value between $85°$ and $86°$.
(vi) Find the tension in the string.

10. A rotary lawn mower uses a piece of light nylon string, with a small metal sphere on the end, to cut the grass. The string is 20 cm in length and the mass of the sphere is 30 g.
(i) Find the tension in the string when the sphere is rotating at 2000 rpm assuming the string is horizonal.
(ii) Explain why it is reasonable to assume that the string is horizontal.
(iii) Find the speed of the sphere when the tension in the string is 80 N.

11. The coefficient of friction between the tyres of a car and the road is 0.8. The mass of the car and its passengers is 800 kg.
(i) Find the maximum frictional force the road can exert on the car and describe what might be happening when this maximum force is acting
(a) at right angles to the line of motion;
(b) along the line of motion.
(ii) What is the maximum speed that the car can travel without skidding on level ground round a circular bend of radius 120 m?

The diagram shows the car, modelled as a particle, now travelling on a banked road at an angle α to the horizontal around a bend of radius 20 cm. The car's speed is such that there is no sideways force (up or down the slope) exerted on its tyres by the road.

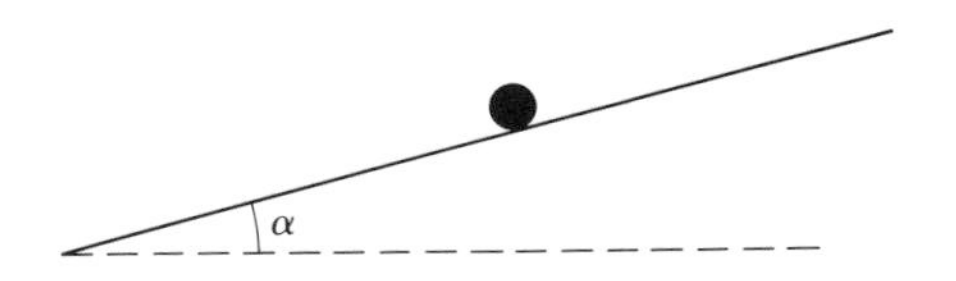

(iii) Draw a diagram showing the weight of the car, the normal reaction of the road on it, and the direction of its acceleration.
(iv) Resolve the forces in the horizontal radial and vertical

directions and write down the horizontal equation of motion and the vertical equilibrium equation.

(v) Show that $\tan\alpha = \dfrac{v^2}{120g}$ where v is the speed of the car in ms^{-1}.

(vi) On this particular bend, vehicles are expected to travel at $15\,\mathrm{ms}^{-1}$. At what angle, α, should the road be banked?

12. Experiments carried out by the police accident investigation department suggest that a typical value for a coefficient of friction between the tyres of a car and a road surface is 0.8.

(i) Using this information, find the maximum safe speed on a circular level motorway slip road of radius 50 m.

(ii) How much faster could cars travel if the road were banked at an angle of 5° to the horizontal?

13. In this question, take g to be $10\,\mathrm{ms}^{-2}$.

An astronaut's training includes periods in a centrifuge.

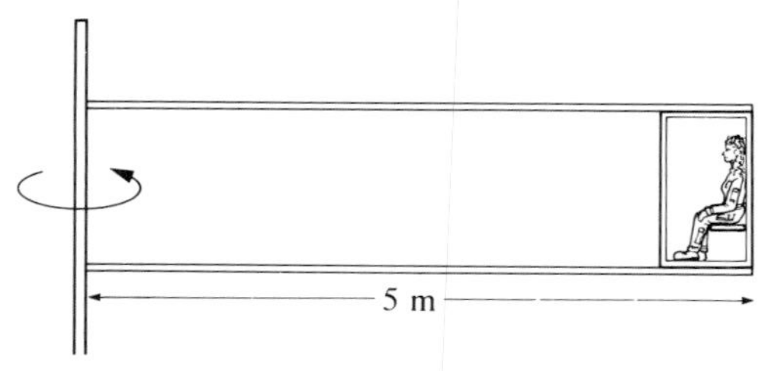

This may be modelled as a cage on the end of a rotating arm of length 5 m.

At a certain time, the arm is rotating at 30 rpm.

(i) Find the angular velocity of the astronaut in $\mathrm{rad\,s}^{-1}$ and her speed in ms^{-1}.

(ii) Show that under these circumstances the astronaut is subject to an acceleration of magnitude about $5g$.

At a later stage in the training, the astronaut blacks out when her acceleration is $9g$.

(iii) Find her angular velocity in rpm when she blacks out.

The training is criticised on the grounds that, in flight, astronauts are subject to linear acceleration rather than rotation and the angular speed is too great. An alternative design is considered in which the astronaut is situated in a carriage driven round a circular railway track. The device must be able to simulate accelerations of up to $10g$ and the carriage can be driven at up to $100\,\mathrm{ms}^{-1}$.

(iv) What should be the radius of the circular railway track?

Circular motion with variable speed

On page 8 you met the general expressions for the acceleration of a particle in circular motion:

- the *transverse component of acceleration* is $r\dot{\omega}$ or $r\ddot{\theta}$ along the tangent;
- the *radial component of acceleration* is $-r\omega^2$ or $-r\dot{\theta}^2$ along the radius.

For circular motion with variable speed, the radial component $-r\dot{\theta}^2$ is the same as that for circular motion with constant speed. The effect of the varying speed appears in the transverse component $r\ddot{\theta}$.

You may remember that on page 3, the speed was found by differentiating $s = r\theta$, with respect to time, to give $\dot{s} = r\dot{\theta}$. The transverse component of acceleration is found by differentiating again: $\ddot{s} = r\ddot{\theta}$.

The symbols $\ddot{\theta}$ and $\dot{\omega}$ denote the rate of change of the angular velocity, or the *angular acceleration*. This quantity is measured in radians per second squared (rad s^{-2}).

In the next two sections you will be studying two types of circular motion with variable speed:
(i) motion with constant angular acceleration and
(ii) unforced motion in a vertical circle.

Circular motion with constant angular acceleration

When the angular acceleration of a body moving in a circle is constant, it is convenient to use a standard notation

$$\ddot{\theta} = \alpha.$$

You can find the angular speed and the angular displacement by integrating this with respect to time:

$$\dot{\theta} = \alpha t + c \quad \text{for some constant } c$$

It is usual to call the initial angular speed ω_0. In this case $\dot{\theta} = \omega_0$ when $t = 0$, so $c = \omega_0$.

This gives $\dot{\theta} = \omega_0 + \alpha t$ or $\omega = \omega_0 + \alpha t$.

Integrating again and assuming that $\theta = 0$ when $t = 0$:

$$\theta = \omega_0 t + \tfrac{1}{2}\alpha t^2$$

These two equations may look familiar to you. They are very much like the equations for motion in a straight line with constant acceleration:

$$v = u + at$$

and

$$s = ut + \tfrac{1}{2}at^2.$$

There is a direct correspondence between the variables.

motion in a straight line with constant acceleration	s	u	v	a
circular motion with constant angular acceleration	θ	ω_0	$\omega(=\dot{\theta})$	α

In fact it can be shown that each equation for motion in a straight line with constant acceleration corresponds to an equation for circular motion with constant angular acceleration.

$$v = u + at \quad \longleftrightarrow \quad \omega = \omega_0 + \alpha t$$

$$s = ut + \tfrac{1}{2}at^2 \quad \longleftrightarrow \quad \theta = \omega_0 t + \tfrac{1}{2}\alpha t^2$$

$$v^2 = u^2 + 2as \quad \longleftrightarrow \quad \omega^2 = \omega_0^2 + 2\alpha\theta$$

$$s = \tfrac{1}{2}(u + v)t \quad \longleftrightarrow \quad \theta = \tfrac{1}{2}(\omega_0 + \omega)t$$

EXAMPLE

Oliver is standing on a playground roundabout at a distance of 2 m from the centre. Imogen pushes the roundabout with constant angular acceleration for 2 seconds and in this time the angular speed increases from $0.3\,\text{rad}\,\text{s}^{-1}$ to $1.5\,\text{rad}\,\text{s}^{-1}$. Find

(i) the angular acceleration of the roundabout;

(ii) the magnitude and direction of the resultant horizontal force acting on Oliver just before Imogen stops pushing the roundabout. (Oliver's mass is 40 kg.)

Solution

(i) Using the standard notation: $\omega_0 = 0.3$, $\omega = 1.5$, $t = 2$. To find α the required equation is:

$$\omega = \omega_0 + \alpha t$$

$$\Rightarrow \quad 1.5 = 0.3 + 2\alpha$$

$$\Rightarrow \quad \alpha = 0.6$$

The angular acceleration is $0.6\,\text{rad}\,\text{s}^{-2}$

(ii)

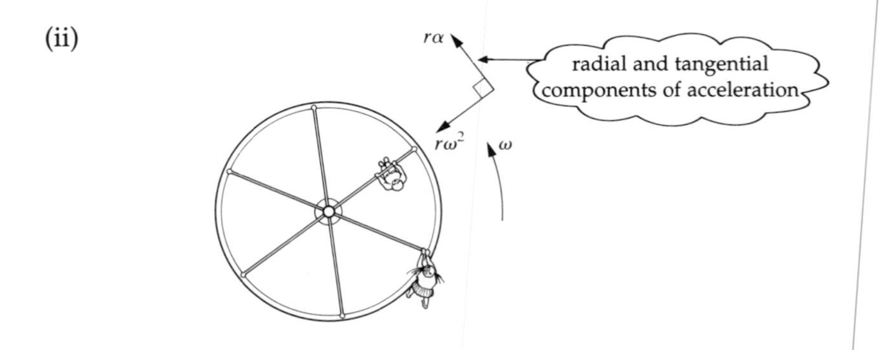

Just before Imogen stops pushing, Oliver's acceleration has two horizontal components.

In the transverse direction $r\ddot{\theta} = r\alpha$

$= 2 \times 0.6$

$= 1.2 \text{ ms}^{-2}$

In the radial direction towards the centre: $r\dot{\theta}^2 = r\omega^2$

$= 2 \times 1.5^2$

$= 4.5 \text{ ms}^{-2}$

The resultant acceleration has two perpendicular components 1.2 and 4.5 ms^{-2}. Its magnitude is $\sqrt{(1.2^2 + 4.5^2)}$ or 4.66 ms^{-2} and it makes an angle arctan $\left(\frac{4.5}{1.2}\right)$ or 75° (to the nearest degree) with the transverse direction shown.

By Newton's Second Law, the resultant horizontal force is given by

$$F = ma$$
$$= 40 \times 4.66 \text{ N}.$$

So the resultant force is 186 N (to 3 significant figures) at 75° to the transverse direction.

Motion in a vertical circle

Figure 1.10 shows the forces acting on a particle of mass m undergoing circular motion in a vertical plane.

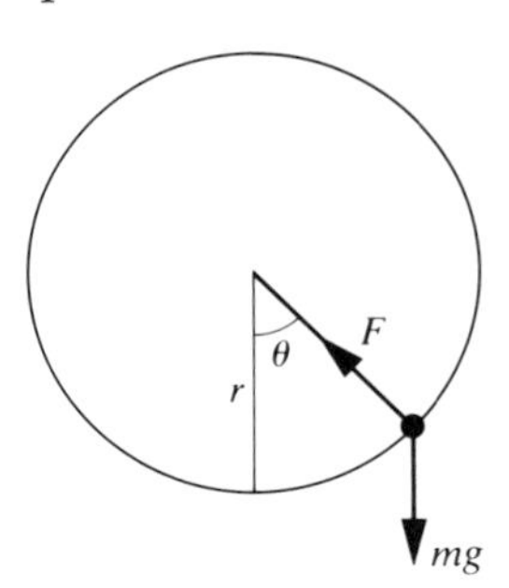

Figure 1.10

For circular motion to take place, there must be a force acting on the particle towards the centre of the circle, as you have seen. This is denoted by F. When the circle is vertical, the force of gravity also acts in this plane, and is therefore relevant to the motion. When the particle is in the position shown in the diagram, Newton's Second Law gives the following equations.

Radial direction (using acceleration $-r\dot{\theta}^2$):

$$mg\cos\theta - F = -mr\dot{\theta}^2$$

Transverse direction (using acceleration $r\ddot{\theta}$):

$$-mg\sin\theta = mr\ddot{\theta}$$

The first of these equations involves the force F which might be the tension in a string or the normal reaction from a surface. This force will vary with θ and so this equation is not helpful in describing how θ varies with time. The second equation, however, does not involve F and may be written as

$$\frac{d^2\theta}{dt^2} = -\frac{g}{r}\sin\theta$$

This differential equation can be solved, using suitable calculus techniques, to obtain an expression for θ in terms of t. The work is beyond the scope of this book but you will meet it in *Differential Equations (Mechanics 4)*.

A different (and at this stage more profitable) approach is to consider the energy of the particle. Since there is no motion in the radial direction, and no friction in the transverse direction, we can apply the principle of conservation of mechanical energy.

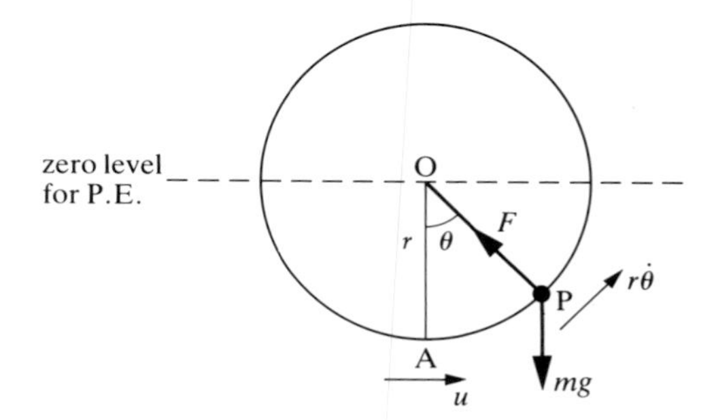

Figure 1.11

Taking u to be the speed of the particle at A, the lowest point of the circle, and the zero level of gravitational potential energy to be that through the centre, O, of the circle as shown in figure 1.11, the total energy at A is

$$\underset{\text{(K.E.)}}{\tfrac{1}{2}mu^2} - \underset{\text{(P.E.)}}{mgr}$$

The total energy at P is

$$\underset{\text{(K.E.)}}{\tfrac{1}{2}m(r\dot{\theta})^2} - \underset{\text{(P.E.)}}{mgr\cos\theta}$$

By the principle of conservation of energy

$$\tfrac{1}{2}m(r\dot{\theta})^2 - mgr\cos\theta = \tfrac{1}{2}mu^2 - mgr$$

$$\Rightarrow \quad r\dot{\theta}^2 = \frac{u^2}{r} - 2g(1 - \cos\theta)$$

This tells you the angular speed, $\dot{\theta}$, of the particle when OP is at any angle θ to OA.

The next two examples show how conservation of energy may be applied to theoretical models of problems involving motion in a vertical circle.

EXAMPLE

A particle of mass $0.03\,\text{kg}$ is attached to the end P of a light rod OP of length $0.5\,\text{m}$ which is free to rotate in a vertical circle with centre O. The particle is set in motion starting at the lowest point of the circle.

The initial speed of the particle is $2\,\text{ms}^{-1}$.

(i) Find the initial kinetic energy of the particle.

(ii) Find an expression for the potential energy gained when the rod has turned through an angle θ.

(iii) Find the value of θ when the particle first comes to rest.

(iv) Find the stress in the rod at this point, stating whether it is a tension or thrust.

(v) Repeat parts (i) to (iv) using an initial speed of $4\,\text{ms}^{-1}$.

(vi) Why is it possible for the first motion (when $v_0 = 2$) to take place if the rod is replaced by a string, but not the second (when $v_0 = 4$)?

Solution

(i)

$$\text{Kinetic energy} = \tfrac{1}{2}mv^2 = \tfrac{1}{2} \times 0.03 \times 2^2 = 0.06$$

The initial kinetic energy is 0.06 joules

(ii) The diagram shows the position of the particle when the rod has rotated through an angle θ.

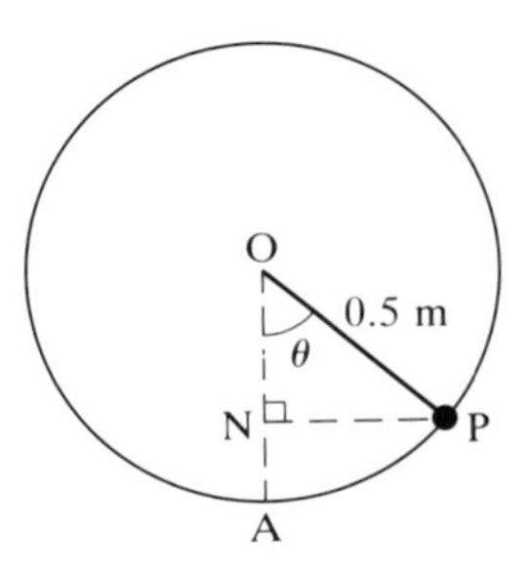

It has risen a distance AN where

$$\begin{aligned}\text{AN} &= \text{OA} - \text{ON} \\ &= 0.5 - 0.5\cos\theta \\ &= 0.5(1 - \cos\theta)\end{aligned}$$

The gain in potential energy at P is therefore

$$0.03g \times 0.5(1 - \cos\theta) = 0.015g(1 - \cos\theta) \text{ joules}$$

(iii) When the particle first comes to rest, the kinetic energy is zero, so by the principle of conservation of energy:

$$0.015g(1 - \cos\theta) = 0.06$$

$$1 - \cos\theta = \frac{0.06}{0.015g}$$

$$= 0.408\ldots$$

$$\Rightarrow \quad \cos\theta = 0.592$$

$$\theta = 0.938 \text{ radians, i.e. about } 53.7°.$$

(iv) The forces acting on the particle are as shown in the diagram.

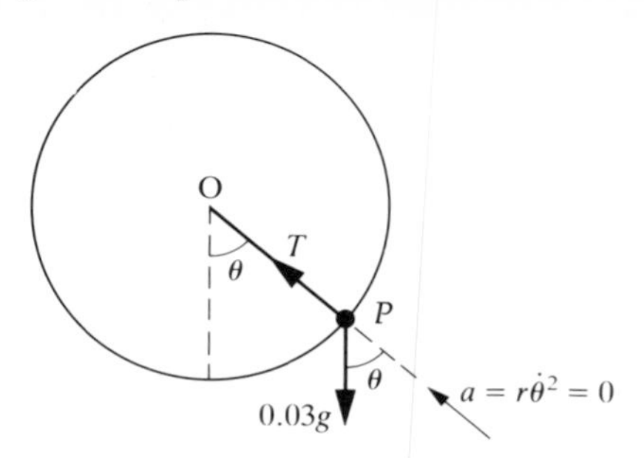

The component of the acceleration towards the centre of the circle is $r\dot{\theta}^2$ which equals zero when the angular speed is zero. Resolving towards the centre:

$$T - 0.03g\cos\theta = 0$$
$$T = 0.174$$

Since this is positive the stress in the rod is a tension. Its magnitude is 0.174 N

(v) The initial K.E. is now

$$\tfrac{1}{2} \times 0.03 \times 4^2 = 0.24 \text{ joules.}$$

The gain in potential energy at P as shown in part (ii).

$$= 0.015\,g(1 - \cos\theta) \text{ joules.}$$

When the particle first comes to rest, the kinetic energy is zero, so by the principle of conservation of energy:

$$0.015g(1 - \cos\theta) = 0.24$$
$$\cos\theta = -0.6327$$
$$\theta = 2.256 \text{ radians, i.e. about } 129°.$$

This time the tension in the rod is

$$T = 0.03g\cos\theta = -0.186$$

The negative tension means that the stress is in fact a thrust of 0.186 N. The diagram illustrates the forces acting in this position.

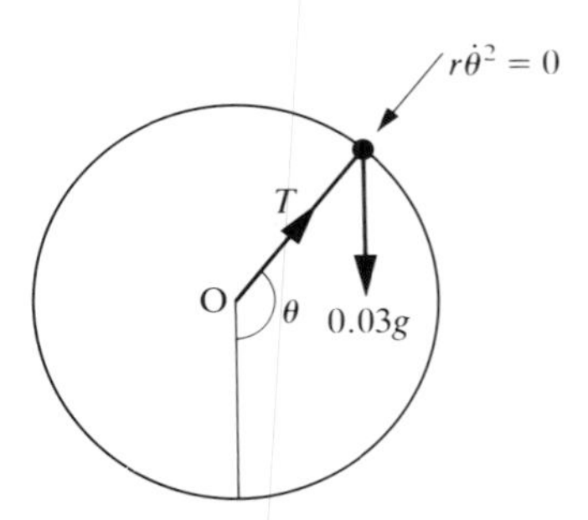

(vi) A string cannot exert a thrust, so although the rod could be replaced by a string in the first case, it would be impossible in the second. In the absence of any radial thrust the particle would leave its circular path at the point where the tension is zero and before reaching the position where the velocity is zero.

EXAMPLE

A bead of mass 0.01 kg is threaded on a smooth circular wire of radius 0.6 m and is set in motion with a speed of u ms^{-1} at the bottom of the circle. This just enables the bead to reach the top of the wire.

(i) Find the value of u.

(ii) What will be the direction of the reaction of the wire on the bead when the bead is at the top of the circle?

Solution

(i) The initial kinetic energy is

$$\tfrac{1}{2}\times 0.01u^2 = 0.005u^2$$

The bead will just reach the top if the speed at the top is zero. If this is the case, its kinetic energy at the top will also be zero.

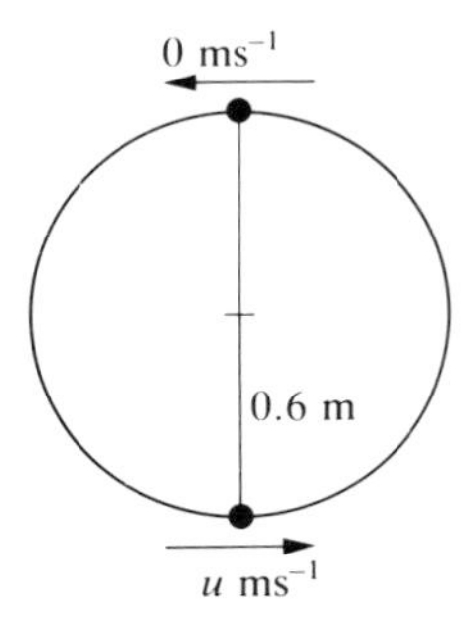

It has then risen a height of 2×0.6 m or 1.2 m, so its gain in potential energy is

$$0.01g\times 1.2 = 0.012g$$

By the principle of conservation of energy,

$$\begin{aligned} \text{loss in K.E.} &= \text{gain in P.E.} \\ 0.005u^2 &= 0.012g \\ u^2 &= 2.4g \\ u &= \sqrt{2.4g} \end{aligned}$$

The initial speed must be 4.85 ms^{-1} to 2 decimal places.

(ii) The reaction of the wire on the bead could be directed either towards the centre of the circle or away from it. The bead has zero angular speed at the top, so the component of its acceleration and therefore the resultant force towards the centre, is zero. The reaction must be outwards, as shown in the diagram, and equal to $0.01g$ newtons.

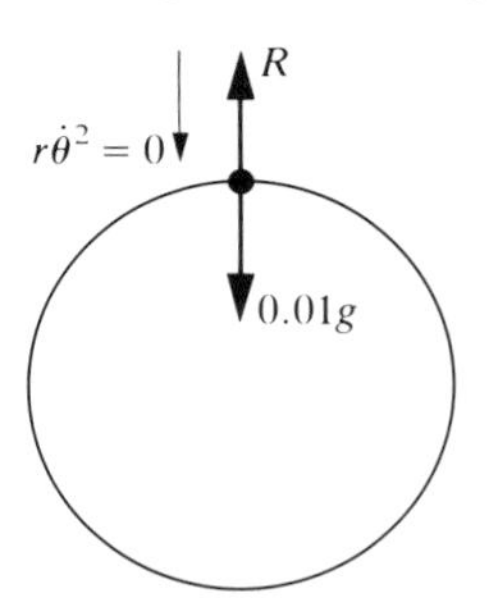

For Discussion

Would this motion be possible if the bead were tied to the end of a string instead of being threaded on a wire?

The breakdown of circular motion

Activity

WARNING: When carrying out this activity, keep well away from other people and breakable objects.

Tie a small object on the end of a piece of strong thread and tie the other end loosely (to minimize friction) round a smooth knitting needle (or a smooth rod with a cork on one end).

Hold the pointed end of the needle and make the object move in a vertical circle as shown in the diagram.

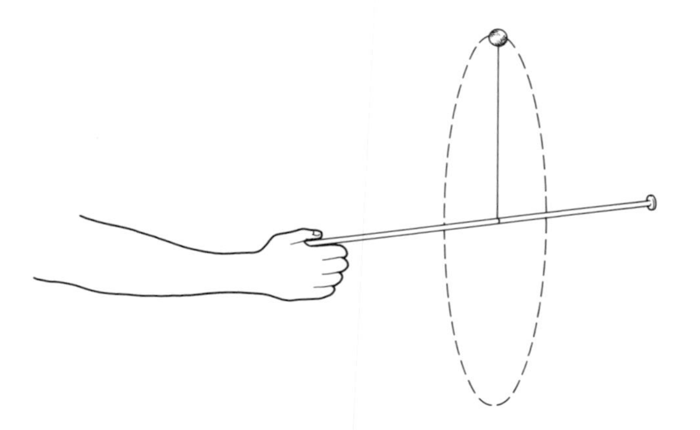

Demonstrate these three types of motion.

1. The object travels in complete circles.
2. The object swings like a pendulum.
3. The object rises above the level of the needle but then fails to complete a full circle.

What would happen if the string broke?
What would happen if a rod were used rather than a string?

The four different types of motion mentioned in the activity are illustrated in figure 1.12.

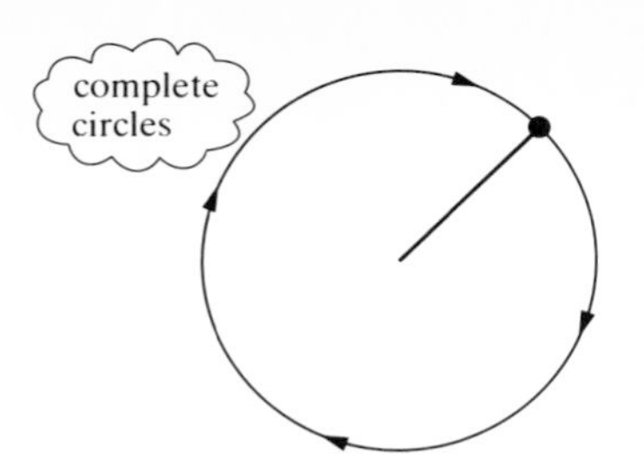

(a) Object oscillates in complete circles

(b) Object oscillates backwards and forwards

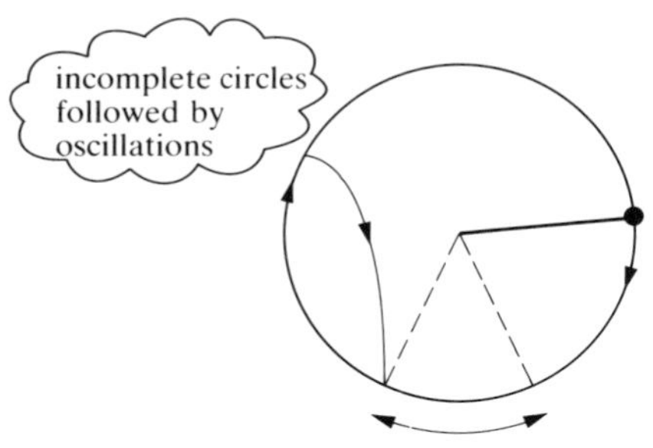

(c) Object leaves circle at some point and falls inwards

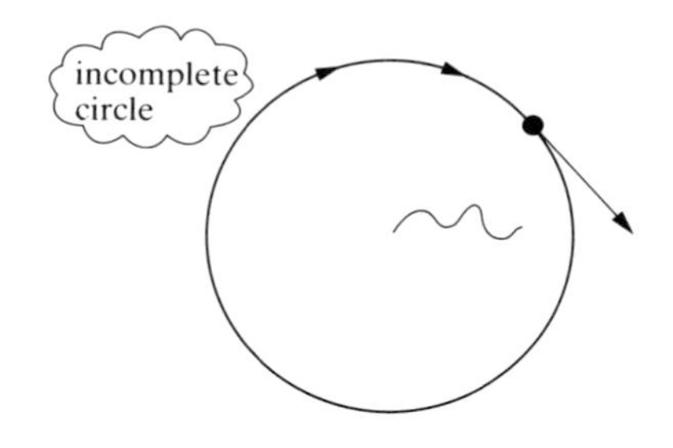

(d) String breaks and object starts to move away along a tangent

Figure 1.12

Modelling the breakdown of circular motion

For what reasons might something depart from motion in a circle? For example, under what conditions will a particle attached to a string and moving in a vertical circle fall out of the circle? Under what conditions will a bicycle travelling over a speed bump with circular cross section leave the road?

A particle on a string

Figure 1.13 shows the forces acting on a particle of mass m attached to a string of length r, rotating with angular speed ω in a vertical circle, centre O.

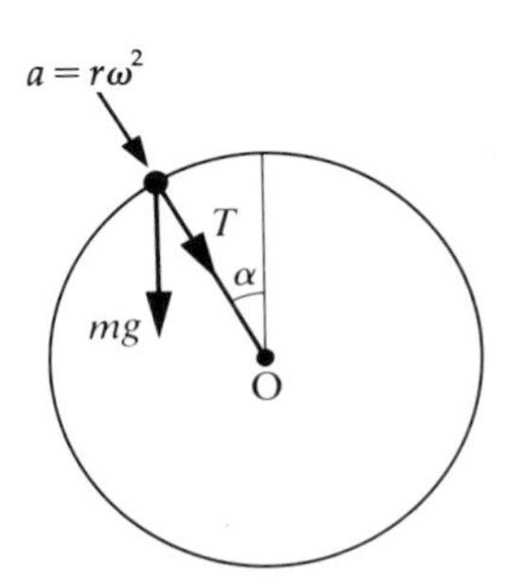

Figure 1.13

There are two forces acting on the particle, its weight mg and the tension T in the string. The acceleration of the particle is $r\omega^2$ towards the centre of the circle.

Applying Newton's Second Law towards the centre gives

$$T + mg\cos\alpha = mr\omega^2 \quad ①$$

where α is the angle shown in the diagram.

While the particle is in circular motion the string is taut, and so $T > 0$. The moment it starts to leave the circle the string goes slack and $T = 0$.

Substituting $T = 0$ in ① gives $\quad mg\cos\alpha = mr\omega^2$

$$\Rightarrow \quad \cos\alpha = \frac{r\omega^2}{g}$$

This equation allows you to find the angle α at which the particle leaves the circle, if it does. The greatest possible value for $\cos\alpha$ is 1, so if $\frac{r\omega^2}{g}$ is greater than 1 throughout the motion, the equation has no solution, and this means that the particle never leaves the circle. Thus the condition for the particle to stay in circular motion is that $\omega^2 \geqslant \frac{g}{r}$ throughout.

In this example, of a particle on the end of a string, ω varies throughout the motion. As you saw on page 28, the value of ω at any instant is given by the energy equation which in this case is

$$\tfrac{1}{2}mr^2\omega^2 + mgr(1 + \cos\alpha) = \tfrac{1}{2}mu^2$$

where u is the speed of the particle at the lowest point.

A particle moving on the inside of a vertical circle

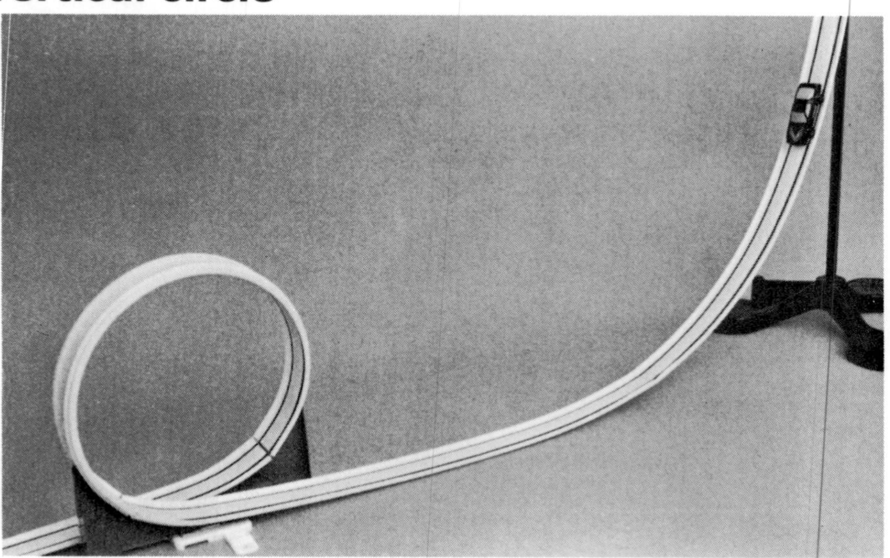

The same analysis applies to a particle sliding around the inside of a smooth circle. The only difference is that in this case the tension, T, is replaced by the normal reaction, R, of the surface on the particle (see figure 1.14). When $R = 0$ the particle leaves the surface.

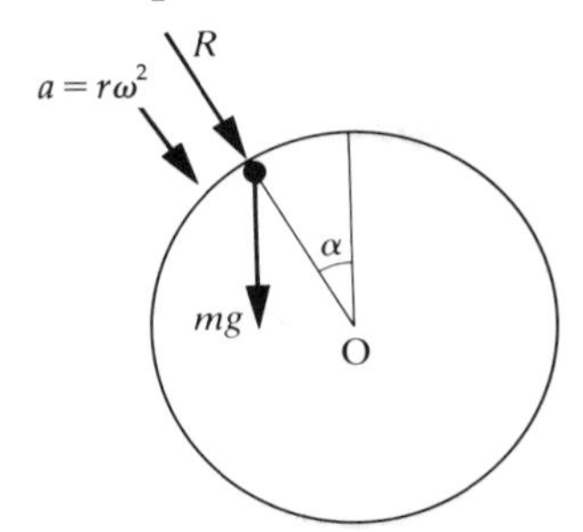

A particle moving on the outside of a vertical circle

The forces acting on a particle moving on the outside of a vertical circle, such as a car going over a hump-backed bridge are the normal reaction, R, acting outwards and the weight of the particle, as shown in figure 1.15.

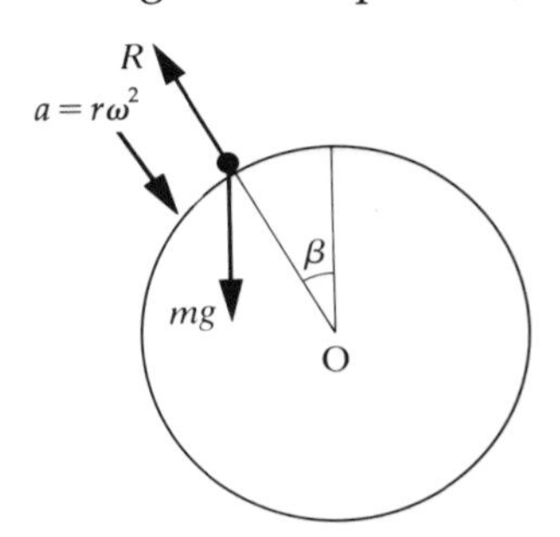

Figure 1.15

Applying Newton's Second Law towards the centre gives

$$mg \cos \beta - R = mr\omega^2 \quad ①$$

where β is the angle shown.

If the normal reaction is zero it means there is no force between the particle and the surface, and so the particle is leaving the surface.

Substituting $R = 0$ in ① gives $\quad mg\cos\beta = mr\omega^2$

$$\Rightarrow \quad \cos \beta = \frac{r\omega^2}{g}$$

For Discussion

The condition for the breakdown of circular motion seems to be the same in the cases of a particle on the end of a string and a particle on the outside of a circle. However everyday experience tells you that circular motion on the end of a string is only possible if the angular speed is large enough whereas a particle will only stay on the outside of a circle if the angular speed is small enough.

How do the conditions $T > 0$ and $R > 0$ explain this difference?

EXAMPLE

Determine whether it is possible for a particle P of mass m kg to be in the position shown in the diagram, moving round a vertical circle of radius 0.5 m with an angular speed of 4 rad s^{-1} when it is

(i) sliding on the outside of a smooth surface;
(ii) sliding on the inside of a smooth surface;
(iii) attached to the end of a string OP;
(iv) threaded on a smooth vertical ring.

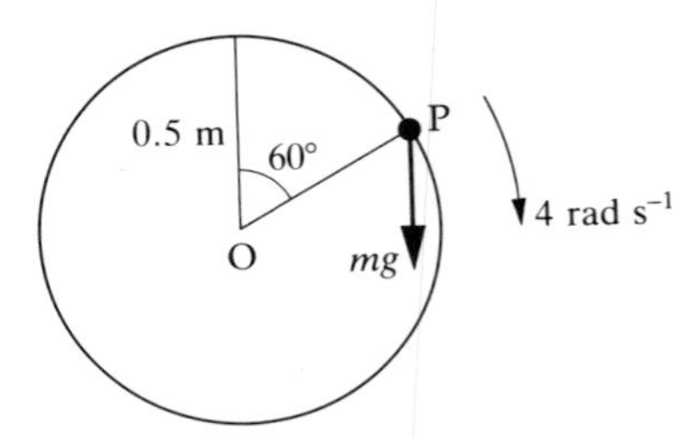

Solution

(i) On the outside of a smooth surface:

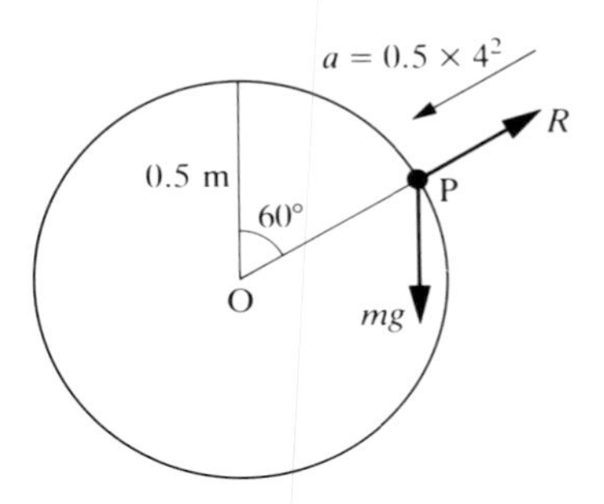

The normal reaction R N of the surface on the particle must be acting outwards, so Newton's Second Law towards the centre gives

$$mg\cos 60° - R = m \times 0.5 \times 4^2$$
$$\Rightarrow \quad R = mg\cos 60° - 8m$$
$$= -3.1m$$

Whatever the mass m, this negative value of R is impossible, so the motion is impossible. The particle will already have left the surface.

(ii) On the inside of a smooth surface:

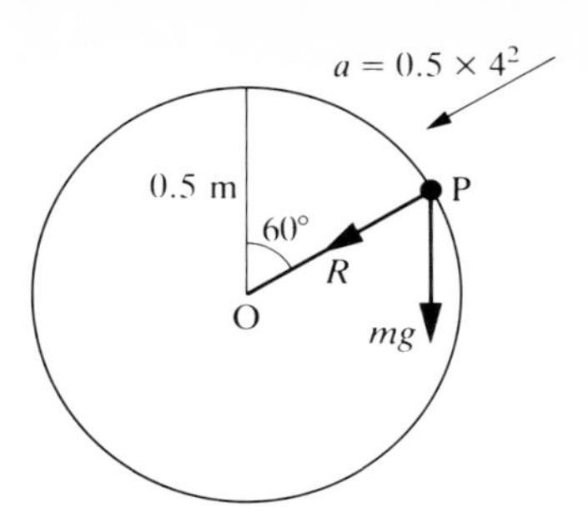

The normal reaction of the surface on the particle will now be acting towards the centre and so

$$R + mg\cos 60° = m \times 0.5 \times 4^2$$
$$\Rightarrow \quad R = +3.1m$$

This is possible.

(iii) Attached to the end of a string:
This example is like (ii) since the tension acts towards the centre, so the motion is possible.

(iv) Threaded on a smooth ring:
If the particle is threaded on a ring the normal reaction can act inwards or outwards so the motion can take place whatever the angular speed. This is also the case when a particle is attached to the centre by a light rod. The rod will exert a tension or a thrust as required.

For Discussion

Which of the situations in the previous example are possible when the angular speed is $3\,\text{rad}\,\text{s}^{-1}$?

EXAMPLE

Eddie, a skier of mass $m\,\text{kg}$, is skiing down a hillside when he reaches a smooth hump in the form of an arc AB of a circle centre O and radius 8 metres as shown in the diagram. O, A and B lie in a vertical plane and OA and OB make angles of $20°$ and $40°$ with the vertical respectively. Eddie's speed at A is $7\,\text{ms}^{-1}$. Determine whether Eddie will lose contact with the ground before reaching the point B.

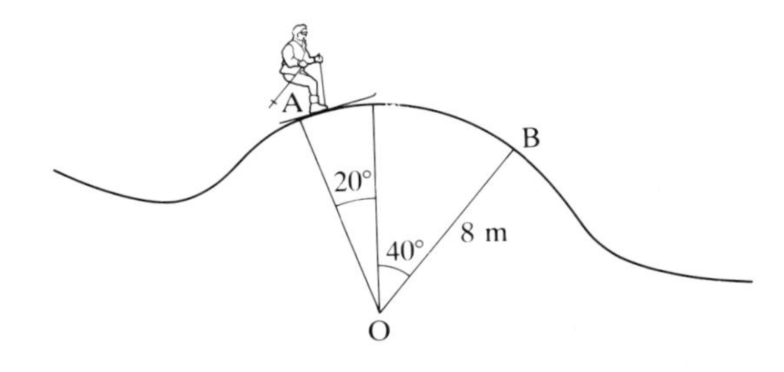

Solution

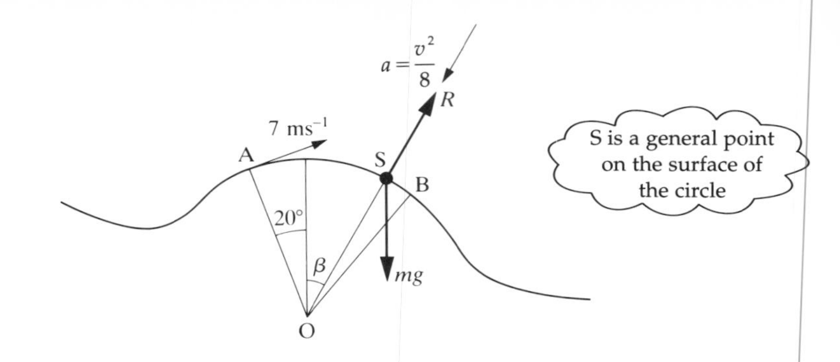

Taking the zero level for potential energy to be a horizontal line through O, the initial energy at A is

$$\tfrac{1}{2}\times m\times 7^2+mg\times 8\cos 20°$$

The energy at any point S is

$$\tfrac{1}{2}mv^2+mg\times 8\cos\beta$$

By the principle of conservation of energy these are equal.

$$\tfrac{1}{2}mv^2+mg\times 8\cos\beta=\tfrac{1}{2}\times m\times 7^2+mg\times 8\cos 20°$$
$$\Rightarrow\quad 0.5v^2+8g\cos\beta=24.5+73.67$$
$$\Rightarrow\quad 0.5v^2=98.17-8g\cos\beta \qquad ①$$

Using Newton's Second Law towards the centre of the circle

$$mg\cos\beta-R=m\frac{v^2}{8}$$

$$\Rightarrow\quad R=m\left(g\cos\beta-\frac{v^2}{8}\right)$$

If Eddie leaves the circle at point S, then

$$R=0\quad\Rightarrow v^2=8g\cos\beta$$

Substituting in ①

$$4g\cos\beta=98.17-8g\cos\beta$$
$$\Rightarrow\quad 12g\cos\beta=98.17$$

This gives $\beta\simeq 33.4°$, which is less than 40°, so Eddie will lose contact with the ground before he reaches the point B.

Exercise 1C

1. A flywheel is initially rotating at 6 radians per second. After 5 minutes it comes to rest. Find

(i) the angular deceleration of the flywheel (assumed constant);

(ii) the total angle through which it turns before coming to rest.

2. A skater spins at 1 revolution per second with her arms out sideways horizontally. She then takes 2 seconds to lower her arms and increases her angular speed to 3 revolutions per second. Find

(i) the angular acceleration in rad s^{-2} (assumed constant);

(ii) the number of complete revolutions she makes during the 2 seconds while she is lowering her arms.

3. A disc slides round a horizontal circular track of radius $2\,\text{m}$. Its initial speed is $8\,\text{m s}^{-1}$ and it has a constant transverse deceleration of $0.001\,\text{m s}^{-2}$.

(i) Calculate the initial angular velocity of the disc.

(ii) Calculate the angular deceleration of the disc.

(iii) Calculate the total angle through which the disc travels before coming to rest.

(iv) Calculate the distance travelled by the disc before it comes to rest.

4. In a model of a discus throw, the discus is rotated in a horizontal circle of radius $0.75\,\text{m}$ with steadily increasing angular speed up to the moment of release. The athlete starts from rest and takes 1.5 seconds to do 1.5 revolutions before releasing the discus. With what speed will it be thrown?

5. The diagram shows the Big Wheel at a fairground. It has radius $3\,\text{m}$. Once it is loaded with passengers it is given a uniform angular acceleration for 20 seconds then runs at uniform angular speed for 60 seconds. It then slows down at a uniform rate over a further 10 seconds. During the main part of the ride, the wheel completes 1 revolution every 4 seconds.

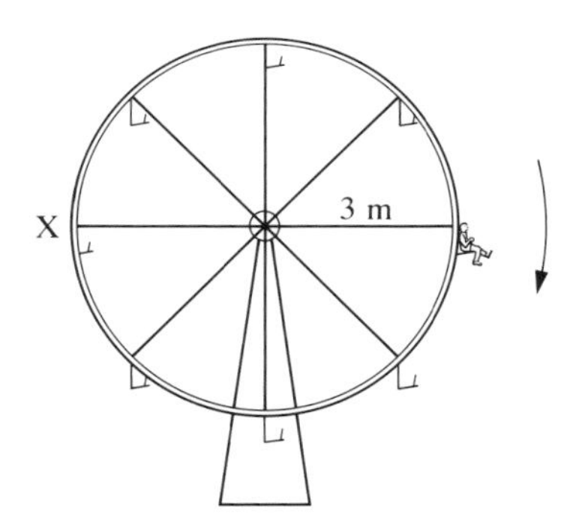

(i) Draw a graph showing the angular speed of the wheel against time and state what information is given by the area under the graph.

(ii) Find the total angle through which a passenger moves, and the distance the passenger travels.

(iii) Find the magnitude of the acceleration of a passenger at the top of the ride when it is travelling at maximum speed. Draw a force diagram to show the forces on a passenger of mass $20\,\text{kg}$ at the highest point of the ride.

Exercise 1C continued

(iv) Describe the acceleration vector of a passenger who is at a point X, when the wheel is half-way through its acceleration phase.

6. The following diagrams show two particles of mass m kg constrained to move in a vertical circle. Their angular speeds and positions are as shown.

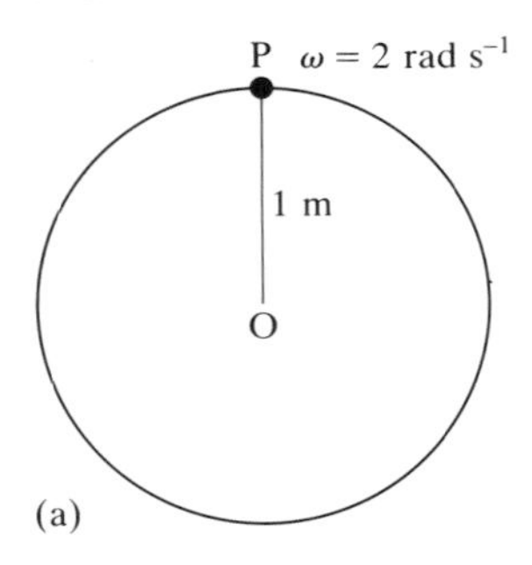

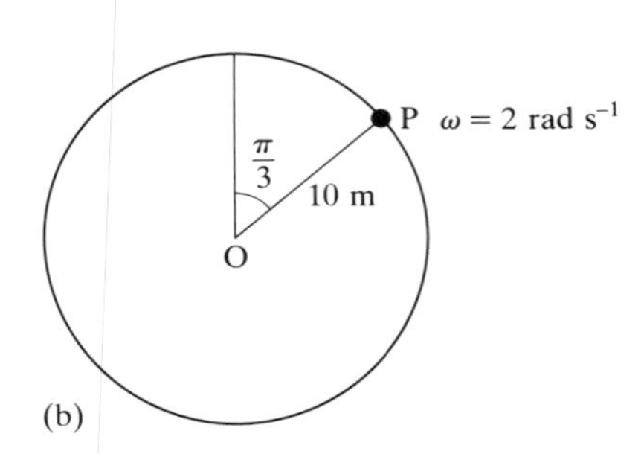

By considering the forces acting on the particle, determine in each case whether it is possible for it to be moving with this speed in this position when it is:

(i) sliding on the outside of a smooth surface;
(ii) sliding on the inside of a smooth surface;
(iii) attached to the end of a string OP;
(iv) threaded on a smooth ring in a vertical plane.

7. Each of the following diagrams shows a particle of mass m which is constrained to move in a vertical circle. Initially it is in the position shown and moving with the given speed.

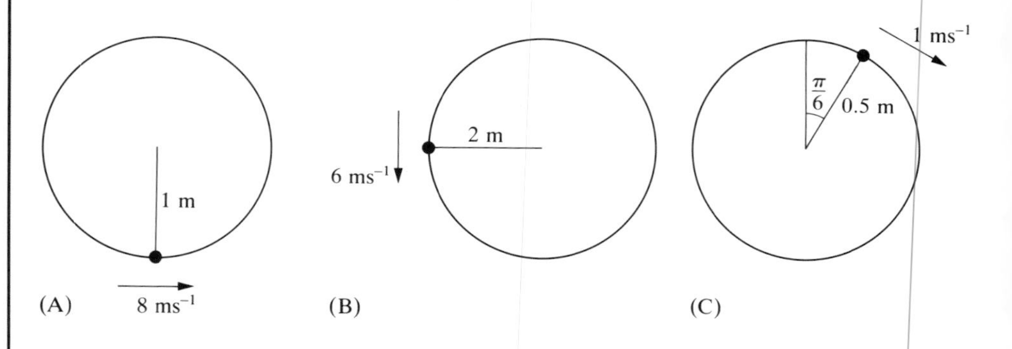

In each case:

(i) Using the horizontal through the centre of the circle as the zero level for potential energy, write down the total initial mechanical energy of the particle.
(ii) Decide whether the particle will make complete revolutions or whether it will come to rest below the highest point. If it makes complete revolutions, determine the speed at the top. Otherwise find the height above the centre when it comes to rest.
(iii) Assuming that the particle is moving under the action of its

weight and a radial force only, find the magnitude and direction of the radial force a) initially and b) when it reaches the top or comes to rest.

8. The diagram shows a model car track. You may assume that all parts of the track lie in the same vertical plane.

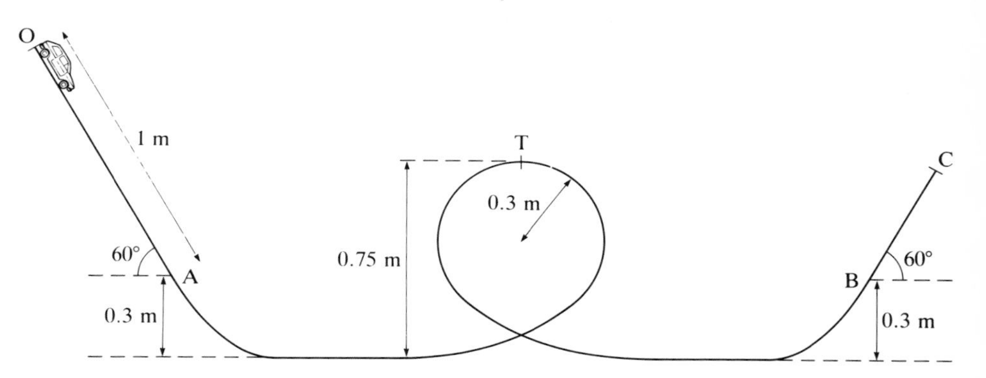

Between A and B the track is symmetrical about a vertical line through T and its length from A to T is $2.1\,\text{m}$. OA and BC are straight and the top of the loop is an arc of radius $0.3\,\text{m}$. For a car of mass $m\,\text{kg}$, there is a frictional resistance of $0.06mg$ N.

The car starts from rest at the point O.

(i) Find the work done against friction between O and T.

(ii) Use the work/energy principle to show that the kinetic energy at T is $0.230mg$ joules.

(iii) By considering circular motion at T, show that the car will move right round the loop in contact with the track.

(iv) The car stops at C before returning. Find the length BC.

(v) Will the car reach T on the return journey?

9. A metal sphere of mass $0.5\,\text{kg}$ is moving in a vertical circle of radius $0.8\,\text{m}$ at the end of a light, inelastic string. At the top of the circle the sphere has speed $3\,\text{ms}^{-1}$

(i) Calculate the gravitational potential energy lost by the sphere when it reaches the bottom of the circle, and hence calculate its speed at this point.

(ii) Calculate the speed of the sphere when the string makes an angle θ with the upward vertical.

(iii) Find the tension in the string when the sphere is (a) at the top and (b) at the bottom point of the circle.

(iv) Draw a diagram showing the forces acting on the sphere when the string makes an angle θ with the upward vertical. Find the tension in the string and the transverse component of the sphere's acceleration at this instant.

10. A glider is travelling horizontally until the pilot executes a loop-the-loop manœuvre as shown in the diagram. The loop may be modelled as a vertical circle. The glider is initially at a height of $700\,\text{m}$, travelling at $30\,\text{ms}^{-1}$. The bottom of the loop is at height $400\,\text{m}$ and the radius of the loop is $100\,\text{m}$.

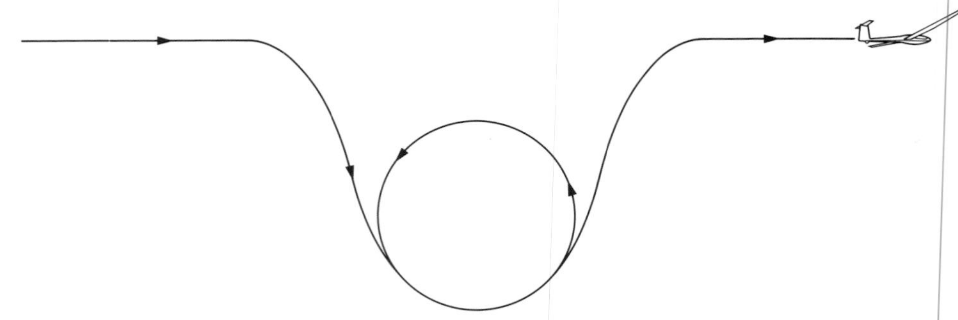

Assuming that mechanical energy is conserved, calculate:

(i) the speed of the glider at the lowest and highest points of the loop;

(ii) the magnitude of the acceleration of the glider at the lowest and highest points of the loop.

The mass of the pilot is $70\,\text{kg}$.

(iii) Draw diagrams to show the forces acting on him at the lowest and highest points of the loop, and state their magnitudes.

(iv) What would happen if he attempted a loop of radius 150 metres, starting from the same lowest point?

(v) What is the maximum radius for a successful loop from this point?

11. As a challenge, a girl is required to swing a bucket of water in a vertical circular arc above her head. If the bucket is not moving fast enough she will get wet. This may be modelled by taking the girl's arm to be $55\,\text{cm}$ long and the bucket $35\,\text{cm}$ from base to handle. The handle is taken to be rigidly attached to the bucket and is held firmly so that her arm is always in line with the centre of the base.

The girl considers three possible depths of water in the bucket: $5\,\text{cm}$, $10\,\text{cm}$ and $15\,\text{cm}$.

(i) For which depth does she need to give the bucket the highest angular speed?

In the event she is forced to have $15\,\text{cm}$ of water in the bucket.

(ii) Find the minimum angular speed at the top of the arc for the water to stay in the bucket.

(iii) Deduce the average angular acceleration the girl must give the bucket, assuming it is at rest at the lowest point.

One girl, doing this for real, ended with the bucket as well as the water hitting her head.

(iv) How could you model this situation?

12. The diagram shows a ride at an amusement park. The loop is, to good approximation, a circle of radius $8\,\text{m}$, in a vertical plane.

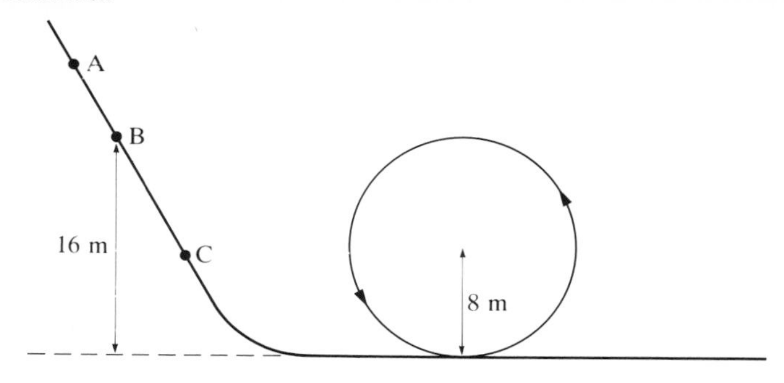

In answering the following questions, you should assume that no energy is lost to forces such as friction and air resistance and the car starts from rest.

(i) Explain why a car which starts just above point B, 16 m above ground level, will not complete the loop.

(ii) For a car to complete the loop successfully it must start at or above point A. What is the height of A?

(iii) On "Kiddiedays" the organisers start the car below point C. Describe what happens to the car and state the maximum height of point C for it to be safe.

13. The diagram illustrates an old road bridge over a river. The road surface follows an approximately circular arc with radius 15 m.

A car is being driven across the bridge, and you should model it as a particle.

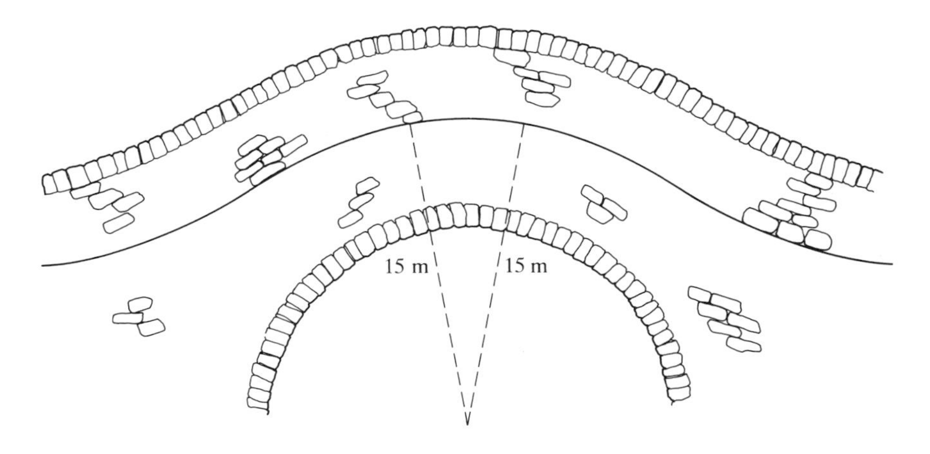

(i) Calculate the greatest constant speed at which it is possible to drive the car across the bridge without it leaving the road, giving your answer both in $\mathrm{ms^{-1}}$ and in mph.

(ii) Comment on the fact that the bridge is old.

(iii) At what point on the bridge would the passengers feel most discomfort?

(iv) How is it possible to improve the design of the bridge?

Investigations

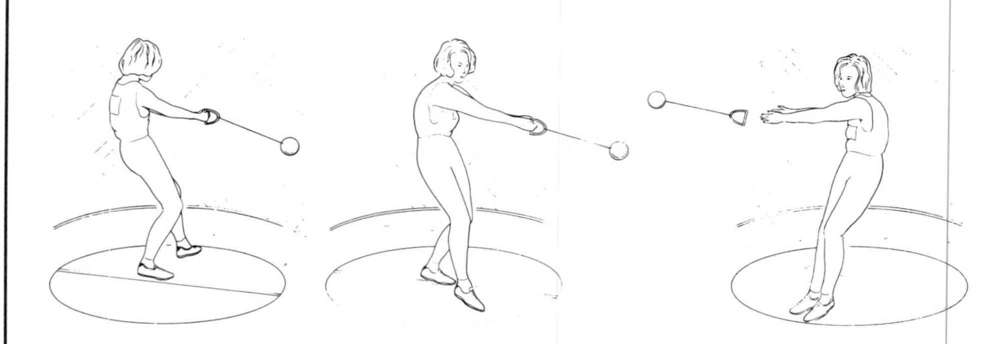

Hammer throwing

Investigate the action of throwing the hammer. Estimate the maximum tension in the wire for a top class athlete. (Data: the hammer is a ball of mass $3\ \mathrm{kg}$ attached to a light wire of length $1.8\ \mathrm{m}$. A throw of $80\ \mathrm{m}$ is world class.)

Loose chippings

Investigate the situation illustrated in this road sign. If you obey the 2 second rule (that you should always be at least 2 seconds behind the car in front of you), and restrict your speed to 20 mph, will you avoid having your windscreen broken?

Why do those taking part in mountain bicycle rallies always go home with mud on their backs?

KEY POINTS

- Position, velocity and acceleration of a particle moving on a circle of radius r.

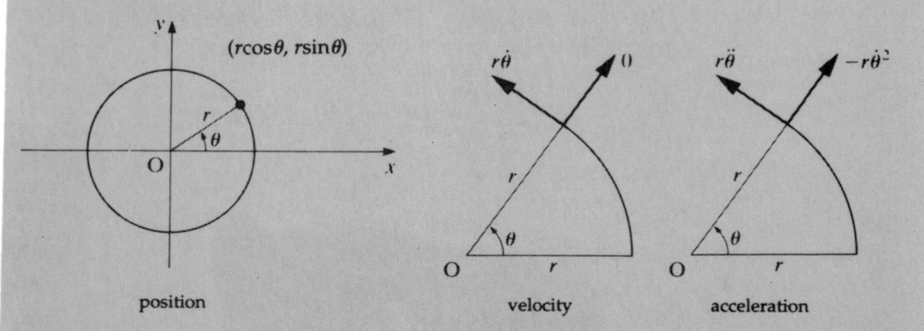

position is $(r\cos\theta, r\sin\theta)$

velocity has transverse component: $v = r\dot{\theta} = r\omega$

radial component: 0

where $\dot{\theta}$ or ω is the angular velocity of the particle.

acceleration has transverse component: $r\ddot{\theta} = r\dot{\omega}$

radial component: $-r\dot{\theta}^2 = -r\omega^2 = -\dfrac{v^2}{r}$

where $\ddot{\theta}$ or $\dot{\omega}$ is the angular acceleration of the particle.

- The forces acting on a particle of mass m in circular motion are given by

transverse component: $mr\dot{\omega} = mr\ddot{\theta}$

radial component: $-\dfrac{mv^2}{r} = -mr\omega^2$

or $-\dfrac{mv^2}{r} = +mr\omega^2$ towards the centre

- Problems involving free motion in a vertical circle can be solved using the conservation of energy principle (P.E. + K.E. = constant) as well as Newton's Second Law.

- The equations for circular motion with constant angular acceleration α are:

$$\omega = \omega_0 + \alpha t \qquad \theta = \omega_0 t + \tfrac{1}{2}\alpha t^2$$
$$\omega^2 = \omega_0^2 + 2\alpha\theta \qquad \theta = \tfrac{1}{2}(\omega_0 + \omega)t$$

where ω_0 is the initial angular speed.

2 Elastic springs and strings

The only way of finding the limits of the possible is by going beyond them into the impossible.

Arthur C. Clarke

The pictures show a person bungee jumping from a bridge.

Bungee jumping is a dangerous sport which originated in the South Sea islands where creepers were used instead of ropes. In the more modern version, a person jumps off a high bridge or crane to which he is attached by an elastic rope round his ankles.

If someone bungee jumping from a bridge wants the excitement of just reaching the surface of the river below, how would you calculate the length of rope required?

The answer to this question clearly depends on the height of the bridge, the mass of the person jumping and the elasticity of the rope. All ropes are elastic to some extent, but it would be extremely dangerous to use an ordinary rope for this sport because the impulse necessary to stop someone falling would involve a very large tension acting in the rope for a short time and this would provide too great a shock to the system. A bungee is a strong elastic rope similar to those used to secure loads on bicycles, cars or lorries, and an essential property for this sport is that it allows the impulse to act over a much longer time and so the rope exerts a smaller force on the jumper.

In this chapter you will be studying some of the properties of elastic springs and strings and will return to the problem of the bungee jumper as a final investigation.

Strings and springs

So far in situations involving strings it has been assumed that the strings do not stretch when they are under tension. Such strings are called *inextensible*. For some materials this is a good assumption, but for others the length of the string increases significantly under tension. Strings and springs which stretch are said to be *elastic*. *Open coiled springs* are springs which can also be compressed.

The length of a string or spring when there is no force applied to it is called its *natural length*. If it is stretched the increase in length is called its *extension* and if a spring is compressed it is said to have a *negative* extension or *compression*.

When stretched, a spring exerts an inward force or tension on whatever is attached to its ends (figure 2.1(a)). When compressed it exerts an outward force or *thrust* on its ends (figure 2.1(b)). An elastic string exerts a tension when stretched, but when slack exerts no force.

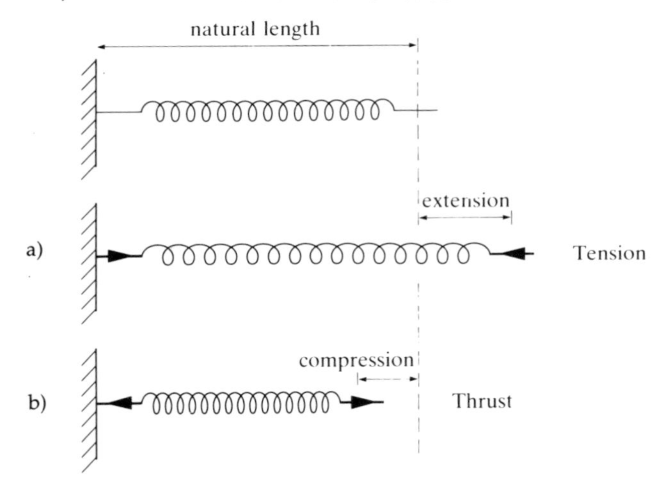

Figure 2.1

Experiment

You will need some elastic, some open coiled springs, some objects of known mass (i.e. weights) and a support stand. Set up the apparatus as shown. Attach one end of an elastic string to the support. You will hang the objects from the other end during the experiment.

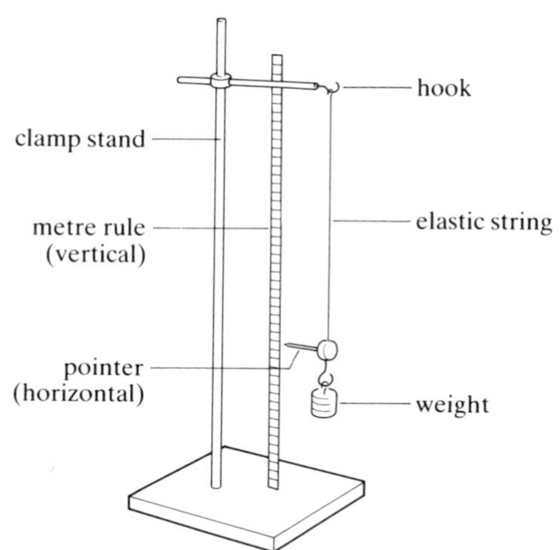

Before doing any experiments, predict the answers to the following questions:

1. How are the extension of the string and the weight of the object hung from it related?
2. If a string of the same material but twice the natural length has the same weight attached, how does the extension change?
3. Does the string return to its original length when unloaded
   (i) if the weight of the object is small?
   (ii) if the weight of the object is large?

Now plot a graph, for each string, of tension, i.e. the weight of the object (vertical axis) against the extension (horizontal axis) to help you to answer these questions.

Design and carry out an experiment which will investigate the relationship between the thrust in an open coiled spring and the decrease in its length.

From your experiments you should have made the following observations.

- Each string or spring returned to its original length once the objects were removed, up to a certain limit.
- The graph of tension or thrust against extension for each string or spring was a straight line for all or part of the data. Strings or springs which exhibit this linear behaviour are said to be *perfectly elastic*.
- The gradient of the linear part of the graph was roughly halved when the string was doubled in length.
- If you kept increasing the weight, the string or spring might have stopped stretching or might have stretched without returning to its original length. In this case the graph would no longer be a straight line: the material had passed its *elastic limit*.
- During your experiment using an open coiled spring you might have found it necessary to prevent the spring from buckling. You may also have found that there came a point when the coils were completely closed and a further decrease in length was impossible.

Hooke's law

In 1678 Robert Hooke formulated a *Rule or law of nature in every springing body* which, for small extensions relative to the length of the string or spring, can be stated as follows:

The tension in an elastic spring or string is proportional to the extension. If a spring is compressed the thrust is proportional to the decrease in length of the spring.

When a string or spring is described as elastic, it means that it is reasonable to apply the modelling assumption that it obeys Hooke's law. A further assumption, that it is light (i.e. has zero mass), is usual and is made in this book.

There are three ways in which Hooke's Law is commonly expressed for a string. Which one you use depends on the extent to which you are interested in the string itself rather than just its overall properties. Denoting the natural length of the string by l_0 and its area of cross section by A, the different forms are as follows.

- $T = \frac{EA}{l_0}x$: In this form E is called *Young's modulus* and is a property of the material out of which the string is formed. This form is commonly used in physics and engineering, subjects in which properties of materials are studied. It is rarely used in mathematics. The SI unit for Young's modulus is $1\,\mathrm{N\,m^{-2}}$.
- $T = \frac{\lambda}{l_0}x$: The constant λ is called the *modulus of elasticity* of the string and will be the same for any string of a given cross section made out of the same material. Many situations require knowledge of the natural length of a string and this form may well be the most appropriate in such cases. The SI unit for modulus of elasticity is 1 N.
- $T = kx$: In this simplest form, k is called the *stiffness* of the string. It is a property of the string as a whole. You may choose to use this form if neither the natural length nor the cross sectional area of the string is relevant to the situation. The SI unit for stiffness is $1\,\mathrm{N\,m^{-1}}$.

Notice that $$k = \frac{\lambda}{l_0} = \frac{EA}{l_0}$$

In this book only the forms using modulus of elasticity and stiffness are used, and these can be applied to springs as well as strings.

EXAMPLE

A light elastic string of natural length 0.7 m and modulus of elasticity 50 N has one end fixed and a particle of mass 1.4 kg attached to the other. The system hangs vertically in equilibrium. Find the extension of the string.

Solution

The forces acting on the particle are the tension T N upwards and the weight $1.4g$ N downwards.

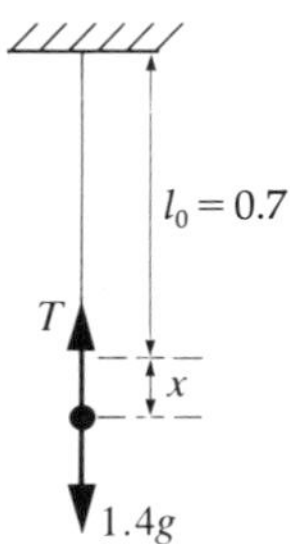

Since the particle is in equilibrium

$$T = 1.4g$$

Using Hooke's law:

$$T = \frac{\lambda}{l_0}x$$

$$\Rightarrow \quad 1.4g = \frac{50}{0.7}x$$

$$\Rightarrow \quad x = \frac{0.7 \times 1.4g}{50}$$

$$= 0.192$$

The extension in the string is 0.19 m correct to 2 significant figures.

EXAMPLE

The mechanism of a set of kitchen scales consists of a light scale pan supported on a spring. When measuring 1.5 kg of flour, the spring is compressed by 7 mm. Find

(i) the stiffness of the spring

(ii) the mass of the heaviest object it can be used to measure if it is impossible to compress it by more than 15 mm.

Solution

(i) The forces on the scale pan with its load of flour are the weight, $1.5g$ N, downwards, and the thrust of the spring T N upwards.

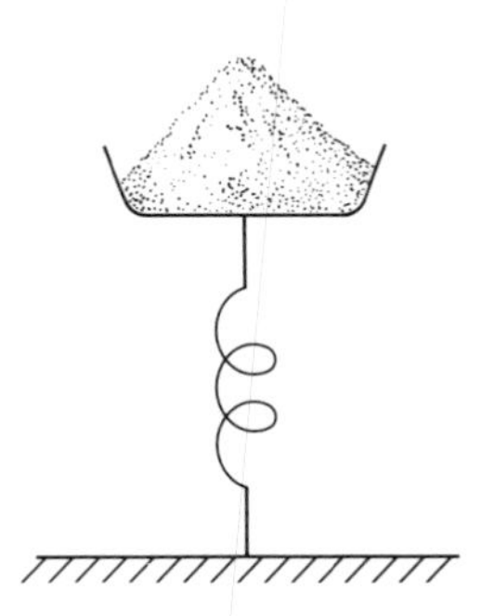

1.5g

T

Since it is in equilibrium

$$T = 1.5g$$

Applying Hooke's law

$T = k \times 0.007$ where k N m^{-1} is the stiffness of the spring.

$$\Rightarrow \quad 1.5g = 0.007k$$

$$\Rightarrow \quad k = 2100$$

The stiffness of the spring is 2100 N m^{-1}.

(ii) Let the mass of the heaviest object be M kg, so the maximum thrust is Mg N. Then Hooke's law gives:

$$Mg = 2100 \times 0.015$$
$$\Rightarrow \quad M = 3.214\ldots$$

The mass of the heaviest object is 3.21 kg, to 3 significant figures.

NOTE *These scales would probably be calibrated to a maximum of 3 kg.*

Exercise 2A

1. A light elastic spring of stiffness k is attached to a ceiling. A block of mass m hangs in equilibrium, attached to the other end of the spring.
   (i) Draw a diagram showing the forces acting on the block.

   The mass of the block is 0.25 kg and the extension of the spring is 0.4 m.
   (ii) Find the value of k in SI units.

   The natural length of the spring is 2 m.
   (iii) Find the modulus of elasticity of the spring.

2. (i) An elastic string has natural length 20 cm. The string is fixed at one end. When a force of 20 N is applied to the other end the string doubles in length. Find the modulus of elasticity.
   (ii) Another elastic string also has natural length 20 cm. When a force of 20 N is applied to each end the string doubles in length. Find the modulus of elasticity.
   (iii) Explain the connection between the answers to (i) and (ii).

3. A light spring has stiffness $0.75\,\mathrm{N\,m^{-1}}$. One end is attached to a ceiling, the other to a particle of weight 0.03 N which hangs in equilibrium below the ceiling. In this situation the length of the spring is 49 cm.
   (i) Find the tension in the spring.
   (ii) Find the extension of the spring.
   (iii) Find the natural length of the spring.

   The particle is removed and replaced with one of weight w N. When this hangs in equilibrium the spring has length 60 cm.
   (iv) What is the value of w?

4. An object of mass 0.5 kg is attached to an elastic string and causes an extension of 8 cm when the system hangs vertically in equilibrium.
   (i) What is the tension in the string?
   (ii) What is the stiffness of the string?
   (iii) What is the mass of an object which causes an extension of 10 cm?

   The modulus of elasticity of the string is 73.5 N.

Exercise 2A continued

(iv) What is the natural length of the string?

5. In this question take the value of g to be $10\ \text{ms}^{-2}$. The diagram shows a spring of natural length 60 cm which is being compressed under the weight of a brick of mass m kg. Smooth supports constrain the brick to move only in the vertical direction.

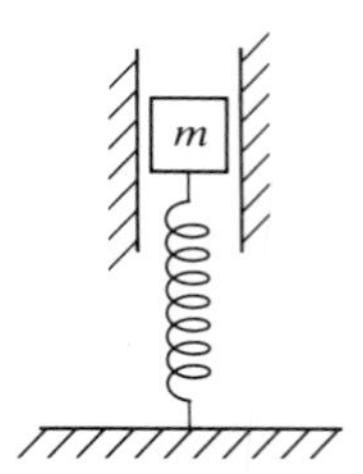

The modulus of elasticity of the spring is 180 N. The system is in equilibrium and the length of the spring is 50 cm. Find:

(i) the thrust in the spring;

(ii) the value of m;

(iii) the stiffness of the spring.

More bricks are piled on.

(iv) Describe the situation when there are seven bricks in total, all identical to the first one.

Using Hooke's law with more than one spring or string

Hooke's law allows you to investigate situations involving two or more springs or strings in various configurations.

EXAMPLE

A particle of mass 0.4 kg is attached to the mid-point of a light elastic string of natural length 1 m and modulus of elasticity λ N. The string is then stretched between a point A in the top of a doorway and a point B which is on the floor 2 m vertically below A.

(i) Find, in terms of λ, the extensions of the two parts of the string

(ii) Calculate their values in the case where $\lambda = 9.8$ N.

(iii) Find the minimum value of λ which will ensure that the lower half of the string is not slack.

Solution

For a question like this it is helpful to draw two diagrams, one showing the relevant natural lengths and extensions, and the other showing the forces acting on the particle.

Since the force of gravity acts downwards on the particle, its equilibrium position will be below the midpoint of AB. This is also shown in the diagram.

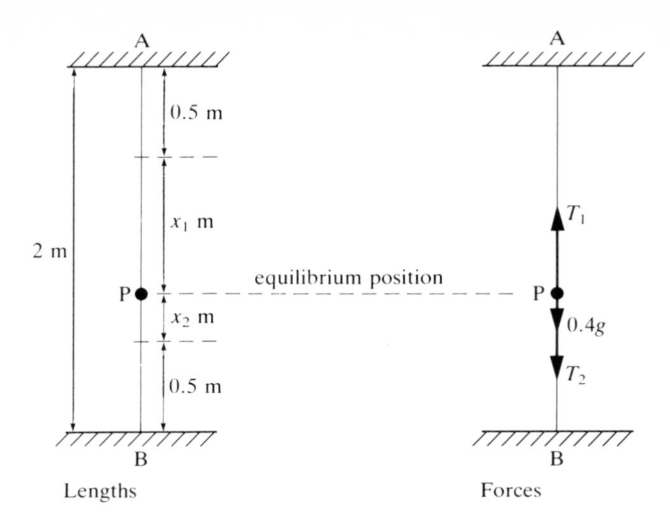

(i) The particle is in equilibrium, so the resultant vertical force acting on it is zero.

Therefore $$T_1 = T_2 + 0.4g \quad ①$$

Hooke's law can be applied to each part of the string.

For AP: $$T_1 = \frac{\lambda}{0.5}x_1 \quad ②$$

For BP: $$T_2 = \frac{\lambda}{0.5}x_2 \quad ③$$

Substituting these expressions in equation ① gives:

$$\frac{\lambda}{0.5}x_1 = \frac{\lambda}{0.5}x_2 + 0.4g$$

$$\Rightarrow \quad \lambda x_1 - \lambda x_2 = 0.5 \times 0.4g$$

$$\Rightarrow \quad x_1 - x_2 = 0.2\frac{g}{\lambda} \quad ④$$

But from the first diagram it can be seen that

$$x_1 + x_2 = 1 \quad ⑤$$

Adding ④ and ⑤ gives:

$$2x_1 = 1 + 0.2\frac{g}{\lambda}$$

$$\Rightarrow \quad x_1 = 0.5 + 0.1\frac{g}{\lambda}$$

Similarly, subtracting ④ from ⑤ gives:

$$x_2 = 0.5 - 0.1\frac{g}{\lambda} \quad ⑥$$

(ii) Since $\lambda = 9.8\,\text{N}$ the extensions are $0.6\,\text{m}$ and $0.4\,\text{m}$.

(iii) The lower part of the string will not become slack providing $x_2 > 0$. It follows from equation ⑥ that:

$$0.5 - 0.1\frac{g}{\lambda} > 0$$

$$\Rightarrow \quad 0.5 > 0.1\frac{g}{\lambda}$$

$$\Rightarrow \quad \lambda > 0.2g$$

The minimum value of λ for which the lower part of the string is not slack is $1.96\,\text{N}$, and in this case BP has zero tension.

HISTORICAL NOTE

Robert Hooke (1635–1703) was the curator of experiments for and later secretary of the Royal Society. He was a contemporary and rival of Newton. In addition to his work on oscillations which led him to the law which has been named after him, Hooke invented the balance spring for watches and worked on problems related to optics and astronomy. His work in astronomy led him to suggest an inverse square law for gravitation. He also designed buildings: there is a church in Milton Keynes of which Hooke is reputedly the architect.

Exercise 2B

1. The diagram shows a uniform plank of weight $120\,\text{N}$ symmetrically suspended in equilibrium by two identical elastic strings, each of natural length $0.8\,\text{m}$ and modulus of elasticity $1200\,\text{N}$.

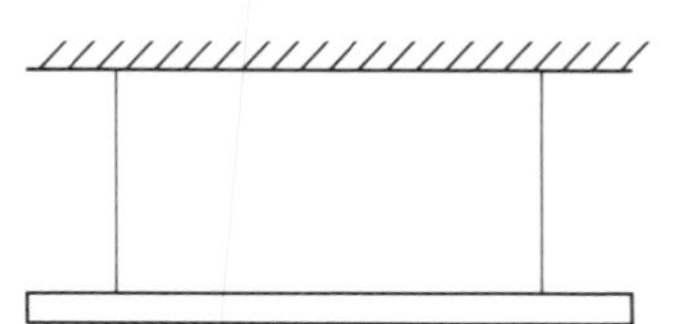

Find:

(i) the tension in each string;

(ii) the extension of each string.

The two strings are replaced by a single string, also of natural length $0.8\,\text{m}$, attached to the middle of the plank. The plank is in the same position.

(iii) Find the modulus of elasticity of this string and comment on its relationship to that of the original strings.

2. The manufacturer of a sports car specifies the coil spring for the front suspension as follows:
a spring of 10 coils with a natural length 0.3 m and a compression 0.1 m when under a load of 4000 N.

(i) Calculate the modulus of elasticity of the spring.

(ii) If the spring were cut into two equal parts, what would be the stiffness of each part?

The weight of a car is 8000 N and half of this weight is taken by two such 10-coil front springs so that each bears a load of 2000 N.

(iii) Find the compression of each spring.

(iv) Two people each of weight 800 N get into the front of the car. How much further are the springs compressed? (Assume that their weight is carried equally by the front springs.)

3. The coach of an impoverished rugby club decides to construct a scrummaging machine as illustrated in the diagrams below. It is to consist of a vertical board, supported in horizontal runners at the top and bottom of each end. The board is held away from the wall by springs, as shown, and the players push the board with their shoulders, against the thrust of the springs.

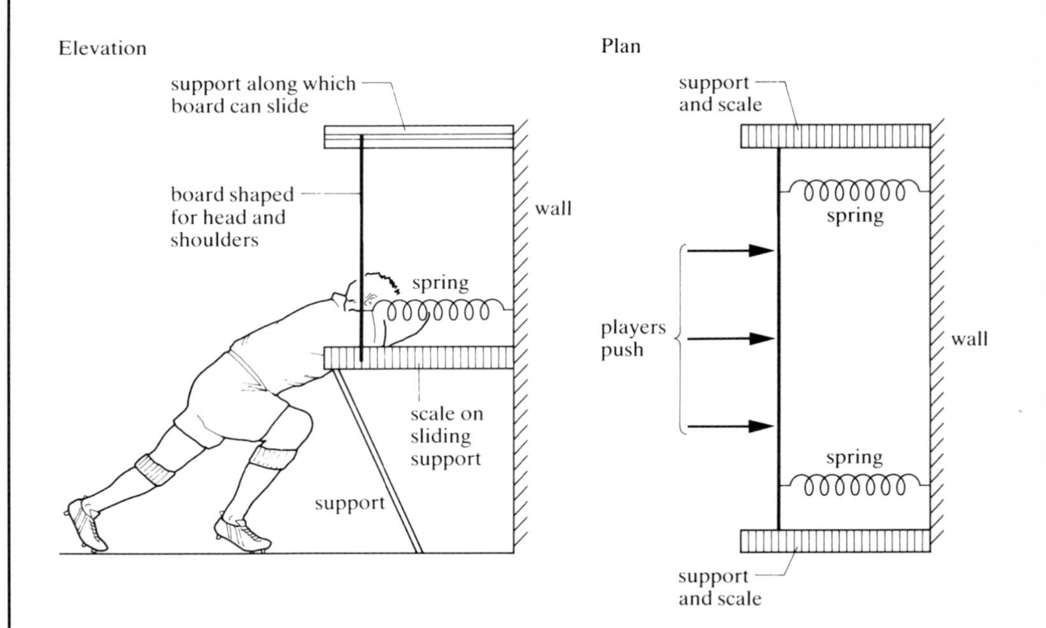

The coach has one spring of length 1.4 m and stiffness 5000 Nm^{-1}, which he cuts into two pieces of equal length.

(i) Find the modulus of elasticity of the original spring.

(ii) Find the modulus of elasticity and the stiffness of each of the half-length springs.

(iii) On one occasion the coach observes that the players compress

Exercise 2B continued

the springs by 20 cm. What total force do they produce in the forward direction?

4. The diagram shows the rear view of a load of weight 300 N in the back of a pick-up truck of width 2 m.

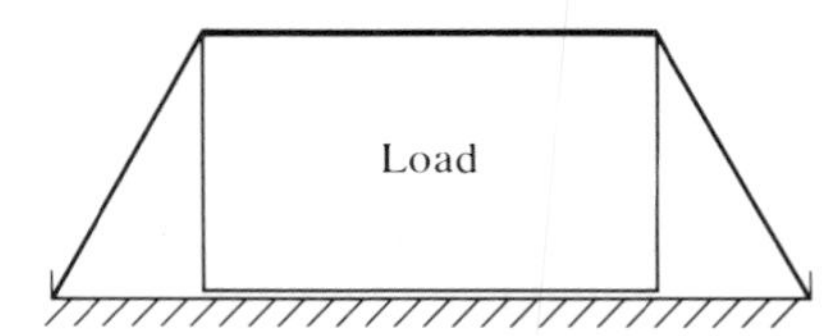

The load is 1.2 m wide, 0.8 m high and is situated centrally on the truck. The coefficient of friction between the load and the truck is 0.4. The load is held down by an elastic rope of natural length 2 m and modulus of elasticity 400 N which may be assumed to pass smoothly over the corners and across the top of the load. The rope is secured at the edges of the truck platform. Find

(i) the tension in the rope;

(ii) the normal reaction of the truck on the load;

(iii) the percentage by which the maximum possible frictional force is increased by using the rope;

(iv) the shortest stopping distance for which the load does not slide, given that the truck is travelling at $30\,\mathrm{ms^{-1}}$ initially. (Assume constant deceleration).

5. Two springs of stiffness k_1 and k_2 are connected in series as shown and a force F is applied to each end.

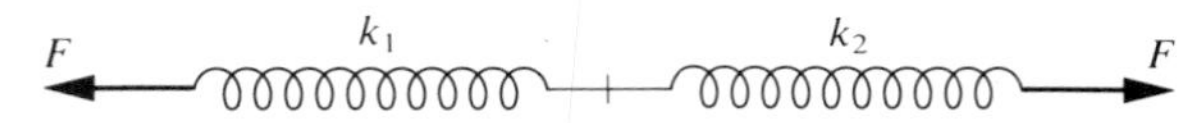

(i) Write down the tension in each spring and find expressions for their extensions in terms of F, k_1 and k_2.

(ii) If these two springs are equivalent to one spring of stiffness k show that:

$$\frac{1}{k} = \frac{1}{k_1} + \frac{1}{k_2}$$

The two springs (which are of the same length) are now connected in parallel, and held so that their extensions are equal.

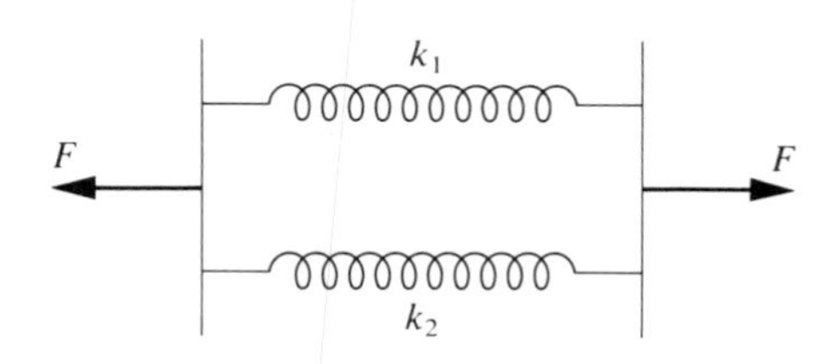

(iii) Show that they are now equivalent to a spring with stiffness $k_1 + k_2$.

6. The diagram shows two light springs, AP and BP, connected at P. The ends A and B are secured firmly and the system is in equilibrium.

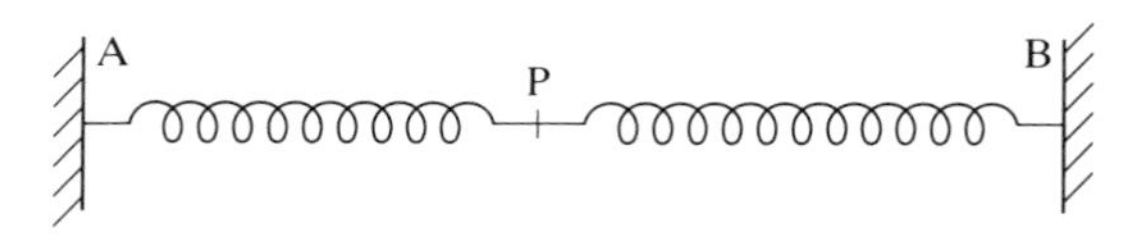

The spring AP has natural length $1\,\text{m}$ and modulus of elasticity $16\,\text{N}$. The spring BP has natural length $1.2\,\text{m}$ and modulus of elasticity $30\,\text{N}$. The distance AB is $2.5\,\text{m}$ and the extension of the spring AP is $x\,\text{m}$.

(i) Write down an expression, in terms of x, for the extension of the spring BP.

(ii) Find expressions, in terms of x, for the tensions in both springs.

(iii) Find the value of x.

7. The diagram shows two light springs, CQ and DQ, connected to a particle of weight $20\,\text{N}$. The ends C and D are secured firmly and the system is in equilibrium, lying in a vertical line.

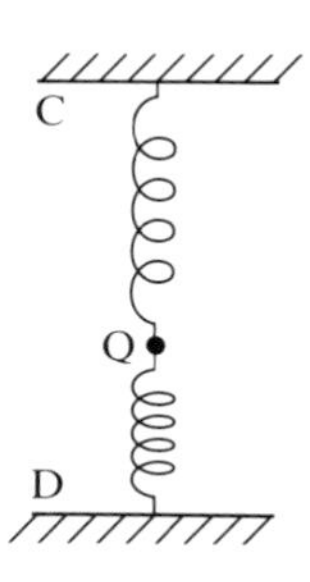

The spring CQ has natural length $0.8\,\text{m}$ and modulus of elasticity $16\,\text{N}$; DQ has natural length $1.2\,\text{m}$ and modulus of elasticity $36\,\text{N}$. The distance CD is $3\,\text{m}$ and QD is $h\,\text{m}$.

(i) Write down expressions, in terms of h, for the extensions of the two springs.

(ii) Find expressions, in terms of h, for the tensions in the two springs.

(iii) Use these results to find the value of h.

(iv) Find the forces the system exerts at C and at D.

8. The diagram (on the next page) shows a block of wood of mass m lying on a plane inclined at an angle α to the horizontal. The block is attached to a fixed peg by means of a light elastic string of natural length l_0 and modulus of elasticity λ; the string lies parallel to the line of greatest slope. The block is in equilibrium.

Exercise 2B continued

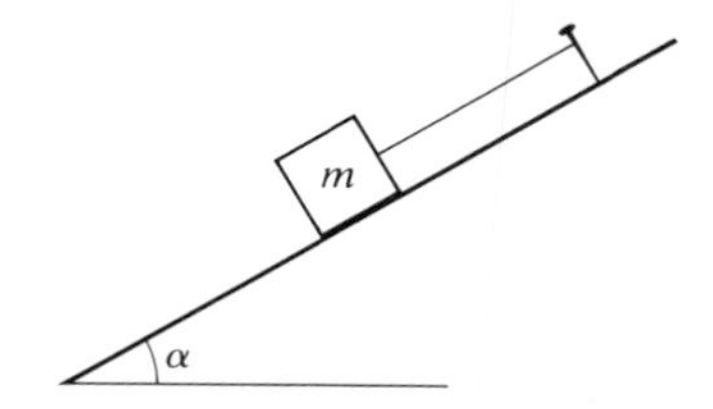

Find the extension of the string in the following cases.
(i) The plane is smooth.
(ii) The coefficient of friction between the plane and the block is μ ($\mu \neq 0$) and the block is about to slide (a) up the plane, (b) down the plane.

9. In this question take g to be $10\ \mathrm{ms}^{-2}$. A baby bouncer consists of a light, inexensible harness attached to a spring. It is suspended in a doorway on the end of a chain so that the baby's feet just touch the floor. The baby can then use its feet to bounce up and down. One such harness, with a spring of natural length $20\ \mathrm{cm}$, is set up for Emily (who has mass $10\ \mathrm{kg}$). Before Emily is put into the harness (i.e. when the spring is not extended) the bottom of the harness is $17\ \mathrm{cm}$ from the ground. When she sits in the harness with her feet off the ground, the bottom of the harness is $15\ \mathrm{cm}$ above the ground.
(i) Find the modulus of elasticity of the spring.

The baby bouncer is also used by Charlotte and its height is not adjusted for her. Charlotte's mass is $12\ \mathrm{kg}$ (the limit recommended for the bouncer).
(ii) What is the extension of the spring when Charlotte sits in the bouncer with her feet off the ground?
(iii) State why Charlotte is unlikely to be seen sitting in this way.
(iv) Find an expression for the reaction, $R\ \mathrm{N}$, between Charlotte's feet and the ground in terms of the height, $h\ \mathrm{cm}$, of the harness above the ground, assuming her to be in equilibrium at the time.

10. A strong elastic band of natural length $1\ \mathrm{m}$ and modulus of elasticity $12\ \mathrm{N}$ is stretched round two pegs P and Q which are in a horizontal line a distance $1\ \mathrm{m}$ apart. A bag of mass $1.5\ \mathrm{kg}$ is hooked onto the band at H and hangs in equilibrium so that PH and QH make angles of θ with the horizontal.
Make the modelling assumptions that the elastic band is light and runs smoothly over the pegs.
(i) Use Hooke's law to show that the tension in the band is $12 \sec \theta$.
(ii) Find the depth of the hook below the horizontal line PQ.
(iii) Is the modelling in this question realistic?

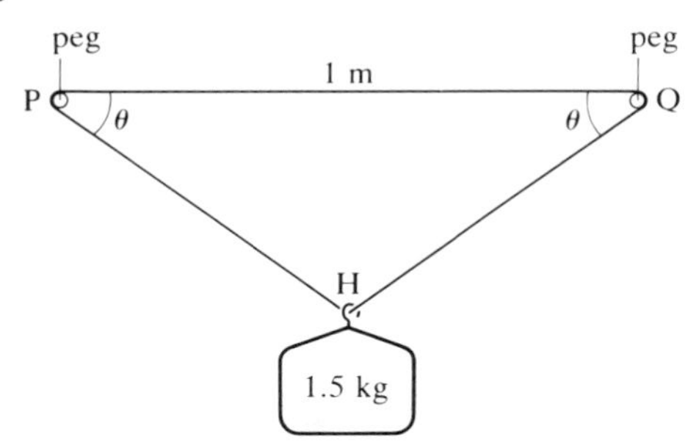

11. A smooth stone of mass 2kg is attached to the end A of an elastic string AB of natural length 1.4m and stiffness 10Nm^{-1}. The end B is fixed to a vertical axle which can rotate smoothly. The stone moves in a circle with centre B on a smooth horizontal table, at constant speed 3ms^{-1}.

(i) Draw a diagram showing the forces acting on the stone.

(ii) Use Newton's Second Law to find the tension in the string.

(iii) Calculate the extension of the string.

Work and energy

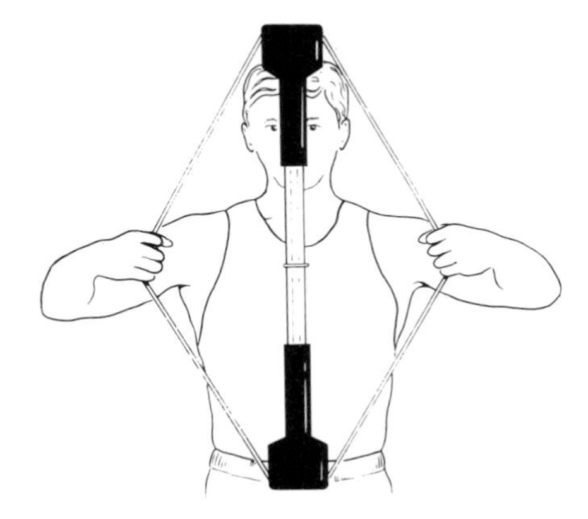

In order to stretch an elastic spring a force must do work on it. In the case of the muscle exerciser, this force is provided by the muscles working against the tension. When the exerciser is pulled at constant speed, at any given time the force F is equal to the tension in the spring; consequently it changes as the spring stretches.

Suppose that one end of the spring is stationary and the extension is x as in figure 2.2. By Hooke's law the tension is given by

$$T = kx,$$

and so

$$F = kx$$

The work done by a constant force F in moving a distance d in its own direction is given by Fd. To find the work done by a variable force the process has to be considered in small stages.

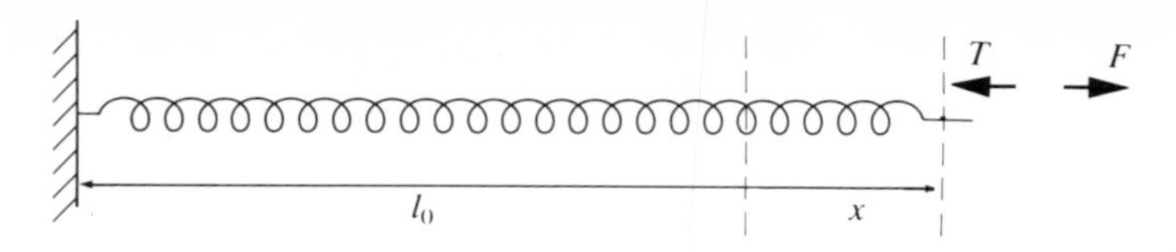

Figure 2.2

Now imagine that the force extends the string a small distance δx. The work done is given by $F\,\delta x = kx\,\delta x$.

The total work done in stretching the spring many small distances is

$$\Sigma F\,\delta x = \Sigma kx\,\delta x$$

In the limit as $\delta x \to 0$, the total work done is:

$$\begin{aligned}\int F\mathrm{d}x &= \int kx\mathrm{d}x \\ &= \tfrac{1}{2}kx^2 + c\end{aligned}$$

When the extension $x = 0$, the work done is zero, so $c = 0$.

The total work done in stretching the spring an extension x from its natural length l_0 is therefore given by:

$$\frac{1}{2}kx^2 \qquad \text{or} \qquad \frac{1}{2}\frac{\lambda}{l_0}x^2$$

The result is the same for the work done in compressing a spring.

Elastic potential energy

The tensions and thrusts in perfectly elastic springs and strings are conservative forces, since any work done against them can be recovered in the form of kinetic energy. A catapult and a Jack-in-a-box use this property.

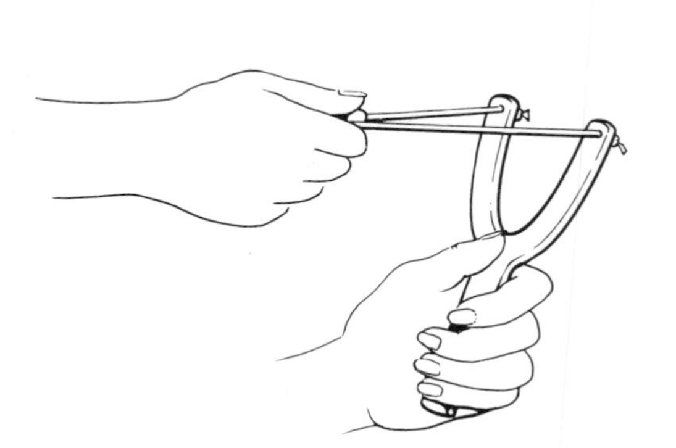

The work done in stretching or compressing a string or spring can therefore be regarded as potential energy. It is known as *elastic potential energy*.

The elastic potential energy stored in a spring which is stretched or compressed by an amount x is

$$\frac{1}{2}kx^2 \qquad \text{or} \qquad \frac{1}{2}\frac{\lambda}{l_0}x^2$$

EXAMPLE

An elastic rope of natural length 0.6 m is extended to a length of 0.8 m. The modulus of elasticity of the rope is 25 N. Find

(i) the elastic potential energy stored in the rope;

(ii) the further energy required to stretch it to a length of 1.65 m over a car roof-rack.

Solution

(i) The extension of the elastic is 0.8 − 0.6 m i.e. 0.2 m.

$$\text{The energy stored in the rope} = \frac{1}{2}\frac{\lambda}{l_0}x^2$$

$$= \frac{25}{2 \times 0.6}(0.2)^2$$

$$= 0.83 \text{ joules (to 2 decimal places).}$$

(ii) The extension of the elastic rope is now $1.65 - 0.6 = 1.05$ m

$$\text{The elastic energy stored in the rope} = \frac{25}{2 \times 0.6}(1.05)^2$$

$$= 22.97$$

The extra energy required to stretch the rope is 22.14 joules (correct to 2 decimal places).

For Discussion

In the example above, the string is stretched so that its extension changes from x_1 to x_2. The work required to do this is

$$\frac{1}{2}kx_2^2 - \frac{1}{2}kx_1^2 = \frac{1}{2}k(x_2^2 - x_1^2) \text{ or } \frac{1}{2}\frac{\lambda}{l_0}(x_2^2 - x_1^2)$$

You can see by using algebra that this expression is **not** the same as $\frac{1}{2}k(x_2 - x_1)^2$, so it is **not** possible to use the extra extension $(x_2 - x_1)$ directly in the energy expression to calculate the extra energy stored in the string. How can you explain this in physical terms?

EXAMPLE

A catapult has prongs which are 16 cm apart and the elastic string is 20 cm long. A marble of mass 70 grams is placed in the centre of the elastic string and pulled back so that the string is just taut. The marble is then pulled back a further 9 cm and the force required to keep it in this position is 60 N. Find:

(i) the stretched length of the string;
(ii) the tension in the string and its stiffness;
(iii) the elastic potential energy stored in the string and the speed which will be given to the marble when the string regains its natural length.

Solution

To solve this problem it is necessary to assume that there is no elasticity in the frame of the catapult, and that the motion takes place in a horizontal plane. In addition, any air resistance is ignored.

In the diagram below, A and B are the ends of the elastic string and M_1 and M_2 are the two positions of the marble (before and after the string is stretched). D is the mid point of AB.

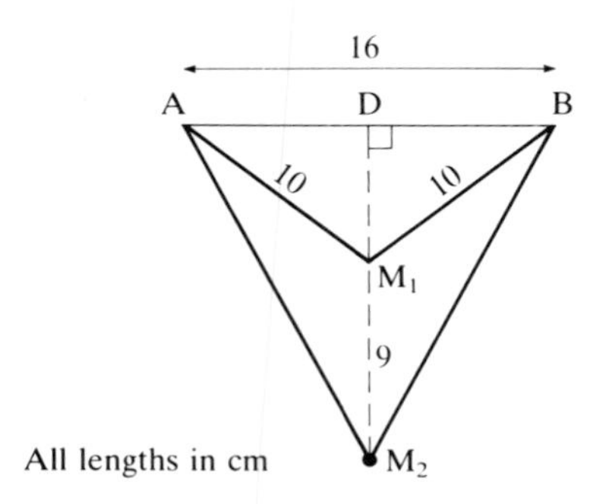

(i) Using Pythagoras' theorem in triangle DBM_1 gives

$$DM_1 = \sqrt{(10^2 - 8^2)} = 6 \text{ cm}.$$

So $DM_2 = 9 + 6 = 15$ cm.

Using Pythagoras' theorem in triangle DBM_2 gives

$$BM_2 = \sqrt{(15^2 + 8^2)} = 17 \text{ cm}$$

The stretched length of the string is 2×17 cm $= 0.34$ m.

(ii) Take the tension in the string to be T newtons.

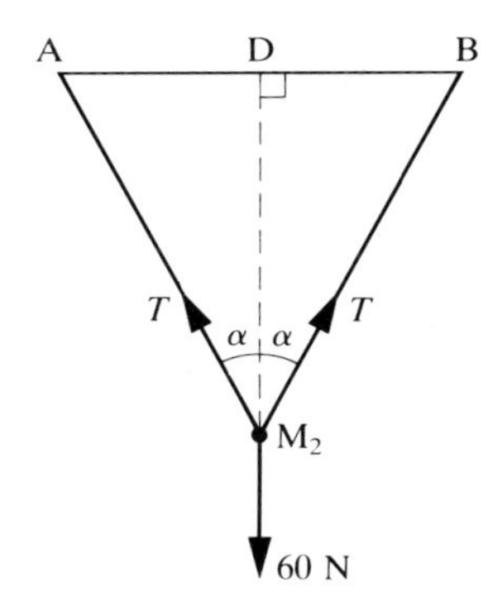

Resolving parallel to M_2D for the equilibrium of the marble gives

$$2T\cos\alpha = 60.$$

Now $\cos\alpha = \dfrac{DM_2}{BM_2} = \dfrac{0.15}{0.17}$

so $T = 34$

The extension of the string is 0.14 m, and by Hooke's law the stiffness k is given by:

$$k = \frac{T}{x}$$

$$= \frac{34}{0.14} = 242.9$$

The stiffness of the string is 243 Nm^{-1} (to 3 significant figures).

(iii) The elastic potential energy stored in the string is

$$\tfrac{1}{2}kx^2 = \tfrac{1}{2} \times 242.9 \times (0.14)^2$$
$$= 2.38 \text{ joules}$$

By the principle of conservation of energy, this is equal to the kinetic energy given to the marble. The mass of the marble is 0.07 kg, so

$$\tfrac{1}{2} \times 0.07v^2 = 2.38$$
$$\Rightarrow \quad v = 8.25 \text{ (to 2 decimal places)}$$

The speed of the marble is 8.25 ms^{-1}.

Exercise 2C

1. An open coiled spring has natural length $0.3\,m$ and stiffness $20\,Nm^{-1}$. Find the elastic potential energy in the spring when

(i) it is extended by $0.1\,m$;
(ii) it is compressed by $0.01\,m$;
(iii) its length is $0.5\,m$;
(iv) its length is $0.3\,m$.

2. A spring has natural length $0.4\,m$ and modulus of elasticity $20\,N$. Find the elastic energy stored in the spring when

(i) it is extended by $0.4\,m$;
(ii) it is compressed by $0.1\,m$;
(iii) its length is $0.2\,m$;
(iv) its length is $0.45\,m$.

3. A pinball machine fires small balls of mass 50 grams by means of a spring and a light plunger. The spring and the ball move in a horizontal plane. The spring has stiffness $600\,Nm^{-1}$ and is compressed by $5\,cm$ to fire a ball.

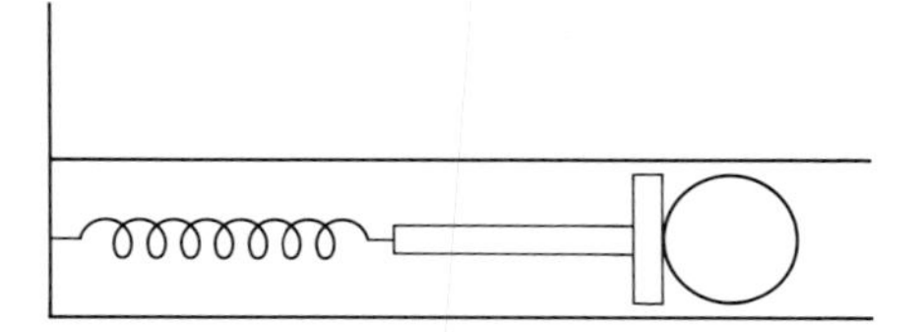

(i) Find the energy stored in the spring immediately before the ball is fired.
(ii) Find the speed of the ball when it is fired.

4. A catapult is made from elastic string with modulus of elasticity $5\,N$. The string is attached to two prongs which are $15\,cm$ apart, and is just taut. A pebble of mass 40 grams is placed in the centre of the string and is pulled back $4\,cm$ and then released in a horizontal direction.

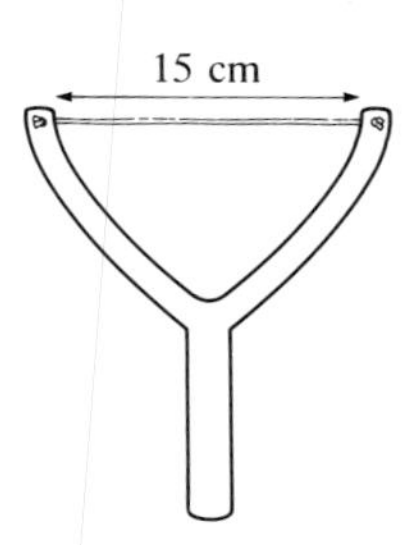

(i) Calculate the work done in stretching the string.
(ii) Calculate the speed of the pebble on leaving the catapult.

5. A simple mathematical model of a railway buffer consists of a horizontal open coiled spring attached to a fixed point. The stiffness of the spring is $10^5\,Nm^{-1}$ and its natural length is $2\,m$.

The buffer is designed to stop a railway truck before the spring is compressed to half its natural length, otherwise the truck will be damaged.

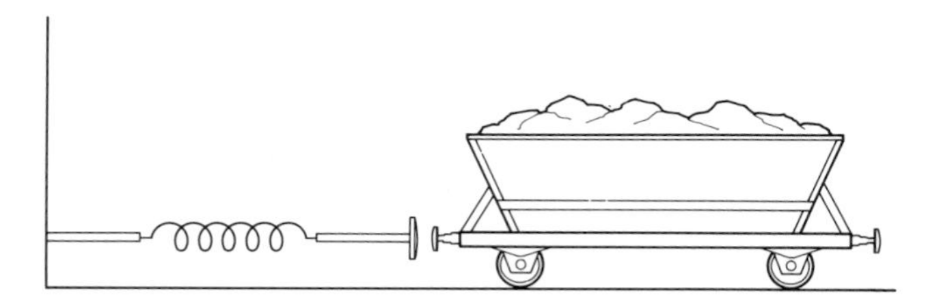

(i) Find the elastic energy stored in the spring when it is half its natural length.

(ii) Find the maximum speed at which a truck of mass 2 tonnes can approach the buffer safely. Neglect any other reasons for loss of energy of the truck.

A truck of mass 2 tonnes approaches the buffer at $5\,\mathrm{ms}^{-1}$.

(iii) Calculate the minimum length of the spring during the subsequent period of contact.

(iv) Find the thrust in the spring and the acceleration of the truck when the spring is at its minimum length.

(v) What happens next?

6. Two identical springs are attached to a sphere of mass $0.5\,\mathrm{kg}$ that rests on a smooth horizontal surface as shown. The other ends of the springs are attached to fixed points A and B.

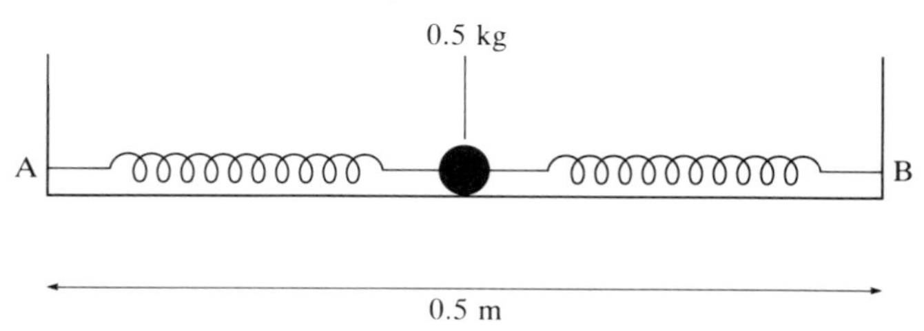

The springs each have stiffness $30\,\mathrm{Nm}^{-1}$ and natural length $25\,\mathrm{cm}$. The sphere is at rest at the mid-point when it is projected with speed $2\,\mathrm{ms}^{-1}$ along the line of the springs towards B. Calculate the length of each spring when the sphere first comes to rest.

Vertical motion

This chapter began with a bungee jumper undergoing vertical motion at the end of an elastic rope. The next example involves a particle in vertical motion at the end of a spring. This, along with the questions in the following exercise, covers the essential work involved in modelling the bungee jumper, which you are then invited to investigate.

EXAMPLE

A particle of mass $0.2\,\text{kg}$ is attached to the end A of a perfectly elastic spring OA which has natural length $0.5\,\text{m}$ and stiffness $40\,\text{N m}^{-1}$. The spring is suspended from O and the particle is pulled down and released from rest when the length of the spring is $0.7\,\text{m}$. In the subsequent motion the extension of the spring is denoted by $x\,\text{m}$.

(i) Write down expressions for the increase in the particle's gravitational potential energy and the decrease in the energy stored in the spring when the extension is x m.

(ii) Hence find an expression for the speed of the particle in terms of x.

(iii) Calculate the length of the spring when the particle is at its highest point.

Solution

(i)

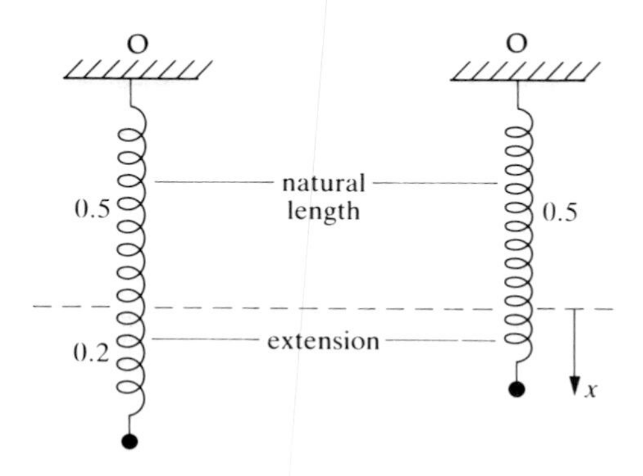

The particle has risen a distance $(0.2 - x)$ m

$$\text{Increase in gravitational P.E.} = mgh$$
$$= 0.2 \times 9.8 \times (0.2 - x)$$

$$\text{Stored energy} = \tfrac{1}{2}kx^2$$
$$= \tfrac{1}{2} \times 40 \times x^2$$
$$\text{Initial stored energy} = \tfrac{1}{2} \times 40 \times 0.2^2$$
$$\text{Decrease in stored energy} = \tfrac{1}{2} \times 40 \times (0.2^2 - x^2)$$

(ii) The initial K.E. is zero

$$\text{Increase in K.E.} = \tfrac{1}{2} \times 0.2 \times \dot{x}^2 - 0$$

$\dot{x}$ is the same as v, the speed in the positive direction

Using the law of conservation of mechanical energy

$$\tfrac{1}{2} \times 0.2\dot{x}^2 + 0.2 \times 9.8 \times (0.2 - x) = \tfrac{1}{2} \times 40 \times (0.2^2 - x^2)$$

and so $v = \dot{x} = \sqrt{[200(0.2^2 - x^2) - 19.6(0.2 - x)]}$

(iii) At the highest point, $v = \dot{x} = 0$,

$$200(0.2^2 - x^2) - 19.6(0.2 - x) = 0$$

$$\Rightarrow \quad x^2 - 0.098x - 0.0204 = 0$$
$$\Rightarrow \quad x = -0.102 \text{ (or 0.2 which was the lowest position)}$$

The negative value of x indicates a compression rather than an extension, so at its highest point the spring has length $(0.5 - 0.102)$ m $= 0.398$ m.

NOTE

For this question it is important that you are dealing with a spring, which still obeys Hooke's law when it contracts, rather than a string which becomes slack.

Exercise 2D

1. A particle of mass $0.2\,\text{kg}$ is attached to one end of a light elastic spring of stiffness $10\,\text{N}\,\text{m}^{-1}$. The system hangs vertically and the particle is released from rest when the spring is at its natural length. The particle comes to rest when it has fallen a distance $h\,\text{m}$.
   (i) Write down an expression in terms of h for the energy stored in the spring when the particle comes to rest at its lowest point.
   (ii) Write down an expression in terms of h for the gravitational potential energy lost by the particle when it comes to rest at its lowest point.
   (iii) Find the value of h.

2. A particle of mass m is attached to one end of a light vertical spring of natural length l_0 and modulus of elasticity $2mg$. The particle is released from rest when the spring is at its natural length. Find, in terms of l_0, the maximum length of the spring in the subsequent motion.

3. A block of mass m is placed on a smooth plane inclined at $30°$ to the horizontal. The block is attached to the top of the plane by a spring of natural length l_0 and modulus λ. The system is released from rest with the spring at its natural length. Find an expression for the maximum length of the spring in the subsequent motion.

4. A particle of mass $0.1\,\text{kg}$ is attached to one end of a spring of natural length $0.3\,\text{m}$ and modulus of elasticity $20\,\text{N}$. The other end is attached to a fixed point and the system hangs vertically. The particle is released from rest when the length of the spring is $0.2\,\text{m}$. In the subsequent motion the extension of the spring is denoted by $x\,\text{m}$.
   (i) Show that $0.05\dot{x}^2 + \dfrac{10}{0.3}(x^2 - 0.1^2) - 0.98(x + 0.1) = 0$
   (ii) Find the maximum value of x.

5. A small apple of mass $0.1\,\text{kg}$ is attached to one end of an elastic string of natural length $25\,\text{cm}$ and modulus of elasticity $5\,\text{N}$. David is asleep under a tree and Sam fixes the free end of the string to the branch of the tree just above David's head. Sam releases the apple level with the branch and it just touches David's head in the subsequent motion. How high above his head is the branch?

Exercise 2D continued

6. A block of mass $0.5\,\text{kg}$ lies on a light scale pan which is supported on a vertical spring of natural length $0.4\,\text{m}$ and modulus of elasticity $40\,\text{N}$. Initially the spring is at its natural length and the block is moving downwards with a speed of $2\,\text{ms}^{-1}$. Gravitational potential energy is measured relative to the initial position and g should be taken to be $10\,\text{ms}^{-2}$

(i) Find the initial mechanical energy of the system.

(ii) Show that the speed $v\,\text{ms}^{-1}$ of the block when the compression of the spring is $x\,\text{m}$ is given by

$$v = 2\sqrt{(1 + 5x - 50x^2)}.$$

(iii) Find the minimum length of the spring during the oscillations.

7. A scale pan of mass $0.5\,\text{kg}$ is suspended from a fixed point by a spring of stiffness $500\,\text{Nm}^{-1}$ and natural length $10\,\text{cm}$.

(i) Calculate the length of the spring when the scale pan is in equilibrium.

(ii) A bag of sugar of mass $1\,\text{kg}$ is gently placed on the pan and the system is released from rest. Find the maximum length of the spring in the subsequent motion.

8. A bungee jump is carried out by a person of mass $m\,\text{kg}$ using an elastic rope which can be taken to obey Hooke's law. It is known that the jump operator does not exceed the total length limit of four times the original length of the rope in any jumps. Prove that the tension in the rope is at most $\frac{8}{3}mg\,\text{N}$.

9. A conical pendulum consists of a bob of mass m attached to an inextensible string of length l. The bob describes a circle of radius r with angular speed ω, and the string makes an angle θ with the vertical as shown.

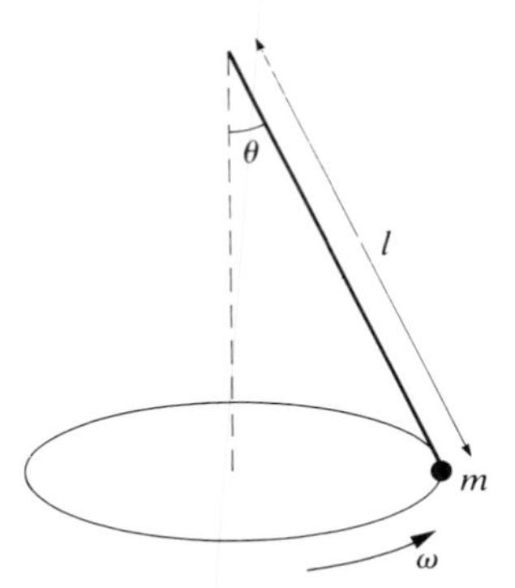

(i) Find an expression for θ in terms of ω, l and g.

The string is replaced with an elastic string of stiffness k and natural length l_0.

(ii) Find an expression for the new value of θ in terms of ω, m, g, l_0 and k.

Investigation

The bungee jumper

(i) *Experiment*

Use a weight to represent the jumper and a piece of elastic for the bungee. Measure l_0 and find the value of λ by suspending the weight in equilibrium. Try to predict the lowest point reached by the weight when it is dropped. Can you estimate a suitable length of elastic for any given weight to fall a standard height?

(ii) *Modelling*

Typical parameters for a mobile crane bungee jump of the type usually done in Britain are shown in the diagram

Height of jump station: 55 m
Bottom safety space: 5 m
Static line length: 5 m (non-elastic straps etc.)
Unstretched elastic rope length: 12 m
Modulus of elasticity: 1000 N

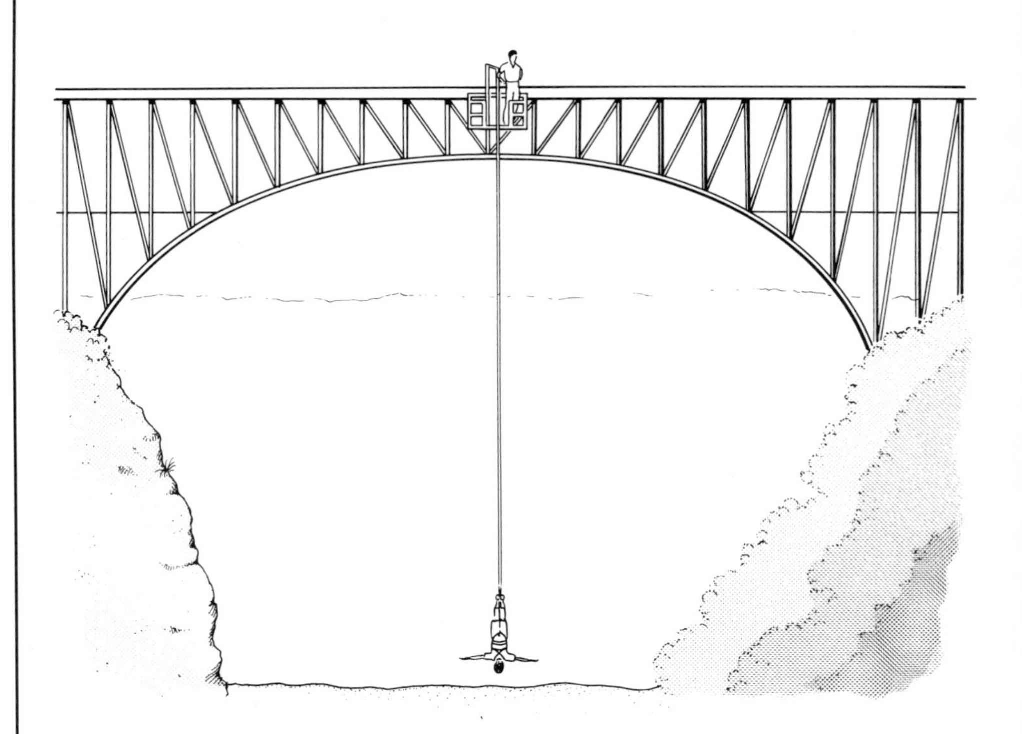

In practice, bungee jumpers usually use a braided rope. The braiding not only keeps the elastic core stretched, it also prevents the rope from stretching too much. As the rope begins to approach its maximum length, the modulus of elasticity gradually increases until "lock out" occurs at maximum extension. This rope then no longer obeys Hooke's law. How would the jump feel different using a braided rope?

Investigation continued

How do you decide on the right length of rope for any jumper? Calculate the maximum deceleration of a jumper using the above information.

KEY POINTS

- **Hooke's law**
  The tension T in an elastic string or spring and its extension x are related by:

$$T = kx \qquad \text{or} \qquad T = \frac{\lambda}{l_0}x$$

  where k is the stiffness, λ is the modulus of elasticity and l_0 is the natural length of the string or spring.
- When a spring is compressed, x is negative and the tension becomes a thrust.
- **Elastic potential energy**
  The elastic potential energy stored in a stretched spring or string, or in a compressed spring, is given by E where:

$$E = \frac{1}{2}kx^2 \qquad \text{or} \qquad E = \frac{1}{2}\frac{\lambda}{l_0}x^2$$

- The tension or thrust in an elastic string or spring is a conservative force and so the elastic potential energy is recoverable.

Modelling oscillations

Backwards and forwards half her length
With a short uneasy motion.

Samuel Coleridge Taylor
The Rime of the Ancient Mariner

What do the guitar, the clock and water ripples have in common?

The guitar, the clock and the water ripples all involve *oscillations* or *vibrations* of particles or bodies.

- In the guitar, the strings vibrate in a controllable way, and the sound box amplifies these vibrations and transmits them as sound waves (vibrations of air molecules).
- In the clock, the pendulum oscillates in the familiar swinging pattern, and this regular motion is used to operate the clock mechanism.
- Ripples are created when the surface of water (or another liquid) is disturbed. The fluid particles vibrate up and down in a regular wave pattern. The same pattern is visible in ocean waves, or in the wakes of boats.

The remarkable thing about all of these vibrations, and many others that occur in natural and man-made systems, is that they are essentially of the same form. If you plot the displacement of a vibrating particle against time for any of these systems, you will obtain a sine wave.

Oscillating motion

The graph in figure 3.1 shows the displacement of an oscillating particle against time.

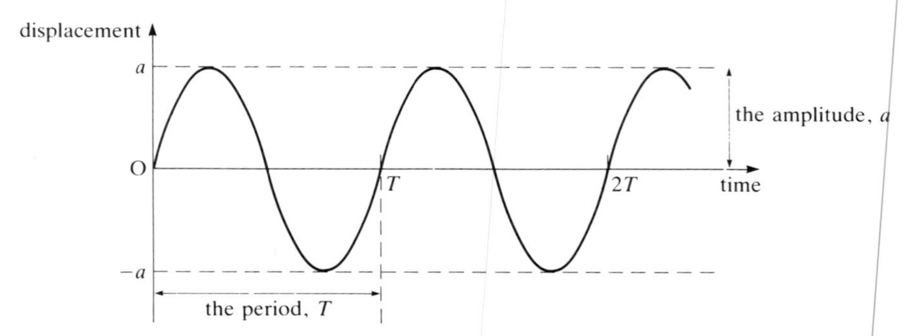

Figure 3.1

From the graph you can see a number of important features of such motion.

- The particle oscillates about a central position, O.
- The particle moves between two points with displacements $+a$ and $-a$. The distance a is called the *amplitude* of the motion.
- The motion repeats itself in a cyclic fashion. The number of cycles per second is called the frequency, and is usually denoted by ν.
- The motion repeats itself after a time T. The time interval T is called the period: it is the time for one complete cycle of the motion.

The terms frequency and period are reciprocally related. For example a period of $\frac{1}{10}$ of a second corresponds to a frequency of 10 cycles per second.

Usually the term *period* is used when describing relatively slow mechanical oscillations and the term *frequency* for faster oscillations or vibrations.

The SI unit for frequency is 1 *hertz* (Hz). One hertz is one cycle per second. The unit is named after Heinrich Rudolph Hertz (1857–94) who was the first person to produce electromagnetic waves artificially. You may be familiar with the use of the megahertz as a measure of the frequency of radio waves. A radio station which is broadcasting on 99.1 megahertz is producing electromagnetic waves with a frequency of 99 100 000 cycles per second.

HISTORICAL NOTE

Pythagoras (572 to 497 BC*) studied the pitch of notes produced by stretched strings in the first recorded acoustic experiments. He showed that recognizable musical intervals are produced by segments of a stretched string when their lengths are in a simple numerical ratio.*

Hooke also did experiments on sound using a strip of brass or card pressed against the teeth of a revolving toothed brass wheel.

In May 1939, shortly before the outbreak of the second world war, an international conference in London unanimously agreed to adopt a frequency of '440 cycles per second for the note "A" in the treble clef.' There are then standard relationships between this and the frequencies of other notes. For many years, before electronic music making devices became readily available, an "A" of this frequency was broadcast regularly on the radio for the use of musicians.

Experiment

Set up a spring–mass oscillator consisting of a weight suspended by a spring as shown in the diagram.

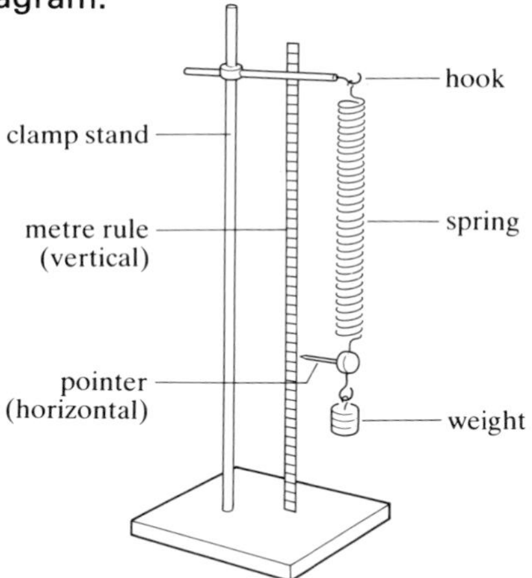

Use the apparatus to investigate whether the period of small oscillations depends on:

(i) the mass of the weight;
(ii) the modulus of elasticity of the spring;
(iii) the natural length of the spring;
(iv) the amplitude of the oscillation.

Keep a record of your conclusions, and see if they are consistent with the theory developed in the next few pages.

Modelling the oscillation of a particle suspended from a spring

The first stage in giving a mathematical description of the motion of the particle in the experiment above is to set up the equation of motion. The work involved is already familiar to you from the previous chapter and from Newton's Second Law, provided you make two modelling assumptions: that the spring is light and that it is perfectly elastic.

The equation of motion of a particle suspended from a spring

The diagram in Figure 3.2 shows a particle suspended from a spring.

Let the natural length of the spring be l_0,
the modulus of elasticity of the spring be λ,
the mass be m,
the extension of the spring when the particle is hanging in equilibrium be e.

Start by finding an expression for e.

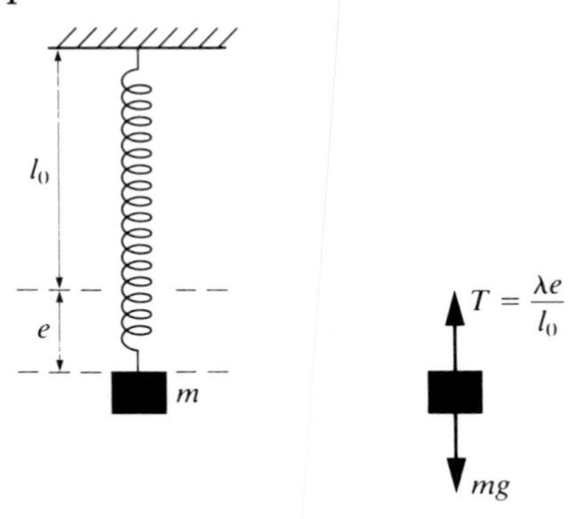

Figure 3.2

The tension T in the spring is given by $T = \frac{\lambda e}{l_0}$ and this is equal to the weight of the particle, mg.

Thus $$\frac{\lambda e}{l_0} = mg \quad (1)$$

Now look at the situation when the spring has an extension x in the downwards direction from the equilibrium position as in figure 3.3. The total extension is now $e + x$.
The particle is not in equilibrium, and its acceleration in the downwards direction is denoted by $\ddot{x}$.

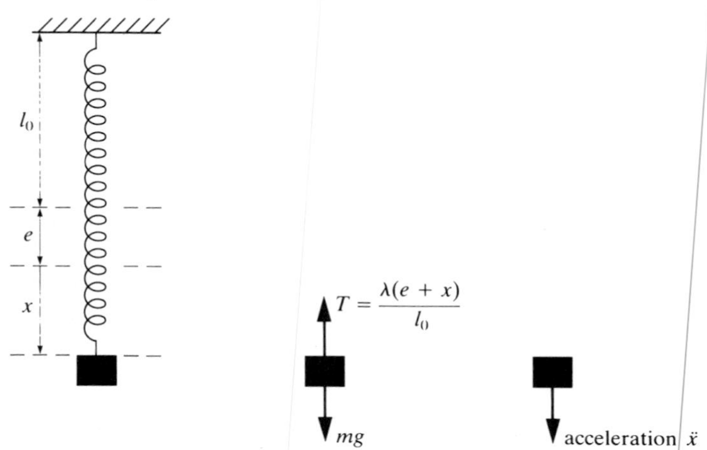

Figure 3.3

There are two forces acting on the particle: the tension in the spring, $T = \frac{\lambda(e+x)}{l_0}$, and its weight, mg.

Applying Newton's Second Law gives

$$m\ddot{x} = mg - \frac{\lambda(e+x)}{l_0}$$

$$= mg - \frac{\lambda e}{l_0} - \frac{\lambda x}{l_0}.$$

However, we have already seen that

$$\frac{\lambda e}{l_0} = mg \qquad \text{(equation ①)},$$

so

$$m\ddot{x} = mg - mg - \frac{\lambda x}{l_0}.$$

This may be simplified to the differential equation

$$\ddot{x} = -\frac{\lambda}{l_0 m}x$$

or

$$\frac{d^2x}{dt^2} = -\frac{\lambda}{l_0 m}x$$

and this is the equation of motion.

The simple harmonic motion equation

At the moment there is no reason why you should be able to take the next step and solve this equation: that is the subject of the rest of this chapter. Before going on to that, it is worth pausing to look at the form of the equation.

In the equation $\ddot{x} = -\frac{\lambda}{l_0 m}x$, λ, l_0 and m are all positive constants, and so may be combined into a single positive constant. By convention this is denoted by ω^2, so that the equation may be written as

$$\ddot{x} = -\omega^2 x.$$

There are many other physical situations which produce an equation of motion of this form. Because the same equation of motion applies to all of them, they all have the same type of oscillating motion, called *simple harmonic motion* (SHM). The guitar strings, the pendulum and the water molecules on page 71 all perform SHM.

Simple harmonic motion is defined by the equation $\ddot{x} = -\omega^2 x$. Remember that in this equation, $\ddot{x}$ means acceleration and x means displacement, so it may be stated in words as follows.

'The acceleration is proportional in magnitude to the displacement from the centre point of the motion, and is directed towards the centre point.'

Notice that it is possible to deduce some features of simple harmonic motion just by looking at the equation, without solving it.

- The value of $\ddot{x}$ is zero when x is zero, so there is zero acceleration at the centre of the oscillation. This means that there must be zero resultant force acting at the centre; it is the *position of equilibrium* of the oscillating system.
- The velocity $\dot{x}$ is not zero at the central position; if the velocity and acceleration are both zero while it remains at the central position, there will be no motion at all.
- The motion is symmetrical about $x = 0$.

It is possible to find a general solution to the equation of motion for SHM by integration (see Mathematical Notes 2, page 159), but it is also possible to demonstrate that the equation is satisfied by certain trigonometrical functions.

EXAMPLE

(i) Show that $x = 10\sin 3t$ satisfies the simple harmonic motion equation

$$\ddot{x} = -9x$$

(ii) Sketch the graph of $x = 10\sin 3t$ and deduce the amplitude, period and frequency of this motion.

(iii) Show that $x = 25\sin 3t$ also satisfies $\ddot{x} = -9x$ and comment on this result.

Solution

(i) $x = 10\sin 3t$

Differentiating with respect to t to find the velocity and acceleration:

$$v = \dot{x} = 30\cos 3t$$

$$\text{and} \quad \ddot{x} = -90\sin 3t$$

$$= -9x$$

This is the SHM equation given in the question.

(ii) The graph of the function is shown below.

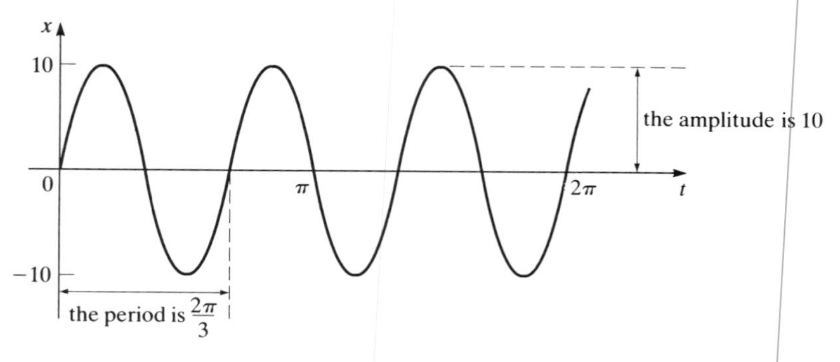

The graph shows that the value of x always lies between -10 and 10, so the amplitude of the motion is 10 units.

Each cycle repeats in the time $\frac{2\pi}{3}$ so the period of the oscillations, denoted by T, is $\frac{2\pi}{3}$. The frequency, f, is $\frac{1}{T}$ or $\frac{3}{2\pi}$.

(iii) Differentiating the function $x = 25\sin 3t$ twice gives:

$$\dot{x} = 75\cos 3t$$
$$\ddot{x} = -225\sin 3t = -9\ (25\sin 3t)$$
$$\Rightarrow \quad \ddot{x} = -9x \text{ as required}$$

This shows that the amplitude of the oscillation is not determined by the differential equation. In this case it can be 10 or 25, or indeed have any other value.

A general form of SHM

Any motion given by an equation of the form $x = a\sin \omega t$ (where a and ω are positive constants) has a similar displacement–time graph and represents SHM with amplitude a. Differentiating this equation gives:

$$v = \dot{x} = a\omega\cos \omega t$$
$$\text{and} \quad \ddot{x} = -a\omega^2\sin \omega t = -\omega^2 x.$$

Figure 3.4 shows how the graphs for x, $\dot{x}$ and $\ddot{x}$ are related.

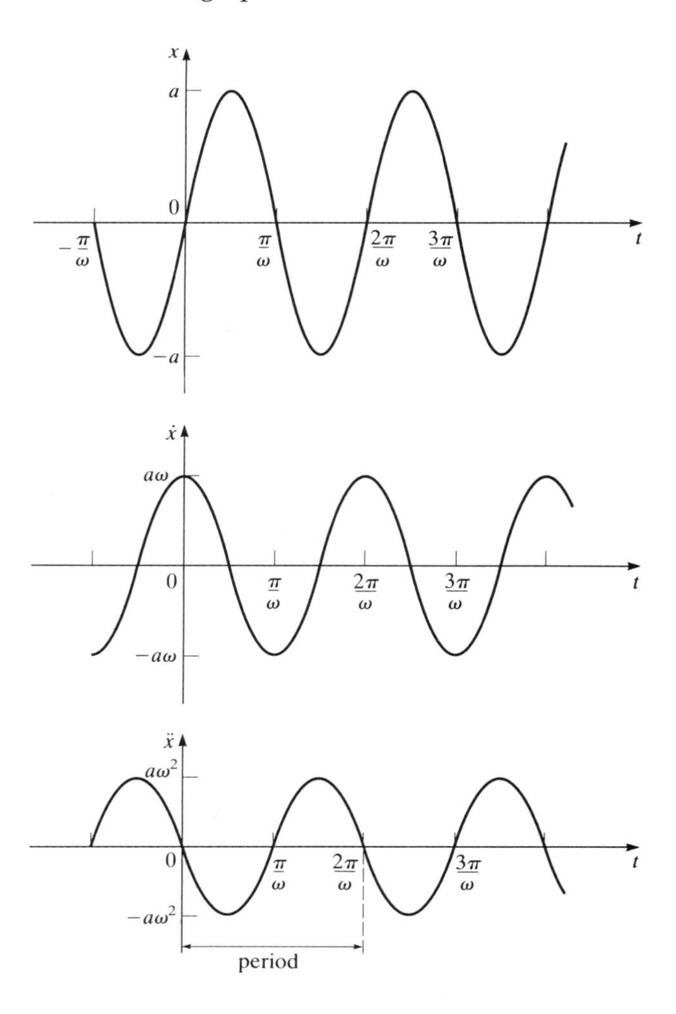

Figure 3.4

The period is shown in figure 3.4, but it can also be found from the equations for x, $\dot{x}$ and $\ddot{x}$. The values of $\sin \omega t$ and $\cos \omega t$, and therefore x, v (or $\dot{x}$) and $\ddot{x}$, all remain unchanged if ωt is increased by 2π and this is the smallest increase in ωt for which this is true.

An increase of 2π in ωt is equivalent to an increase of $\dfrac{2\pi}{\omega}$ in t, and so this is the period of the motion.

In general, $x = a\sin \omega t$ represents SHM with the following properties.

- The amplitude is $a(>0)$
- The period T is $\dfrac{2\pi}{\omega}$
- The frequency $f = \dfrac{1}{T} = \dfrac{\omega}{2\pi}$
- $x = 0$ when $t = 0$

Activity

Show that the equation $x = a\cos \omega t$ also satisfies the definition of SHM. Draw sketches of the graphs of $x = a\sin \omega t$ and $x = a\cos \omega t$ and compare the properties of the motions represented by them.

Phase difference

The graph of $x = a\cos \omega t$ is a translation of the graph of $x = a\sin \omega t$ along the t axis by a displacement of $-\dfrac{\pi}{2}$ (or $\dfrac{3\pi}{2}$ or $\dfrac{7\pi}{2}$ etc.). Both equations describe SHM but their oscillations are at a different stage in the cycle when $t = 0$. For $x = a\cos \omega t$, $x = a$ when $t = 0$. They are said to have a *phase difference* of $-\dfrac{\pi}{2}$ (or $\dfrac{3\pi}{2}$ etc.)

Angular frequency

In these mathematical models ωt is an angle; ω is sometimes called the *angular frequency*. It is important to note that, when the model is applied to an oscillating system, ω is a constant which depends on the properties of the system, such as the stiffness of a spring or the length of a pendulum: ω rarely involves a physical angle and, in particular, ω is *not* the angular velocity of the pendulum.

Relationships between acceleration, velocity and displacement

Provided t is measured from an instant when the particle is at the centre and moving in the positive direction, simple harmonic motion is described by the following equations.

Displacement: $x = a\sin \omega t$ ①
Velocity: $v = \dot{x} = a\omega\cos \omega t$ ②
Acceleration: $\ddot{x} = -a\omega^2\sin \omega t$ ③

It can be seen from equations ① and ③ that

$$\ddot{x} = -\omega^2 x \qquad ④$$

Squaring equation ② gives:

$$v^2 = \dot{x}^2 = a^2\omega^2\cos^2\omega t$$
$$= \omega^2(a^2 - a^2\sin^2 \omega t).$$

Therefore $v^2 = \omega^2(a^2 - x^2)$ ⑤

Equations ④ and ⑤ are also true if $x = a\cos \omega t$, and many properties of SHM can be derived from these two equations alone.

From equation ⑤ you can see that:

- at the extreme points, when $x = \pm a$, the velocity, v or $\dot{x}$ is zero;
- x must lie between $-a$ and a, otherwise v^2 would be negative;
- v has a maximum magnitude of ωa when $x = 0$.

Equation ④ indicates that:

- the acceleration $\ddot{x}$ is zero when $x = 0$;
- $\ddot{x}$ has a maximum magnitude of $\omega^2 a$ when $x = \pm a$, i.e. at the extreme points.

Figure 3.5 illustrates these results for a particle oscillating with SHM between two points C and D. The point O is the equilibrium position, and CD = $2a$.

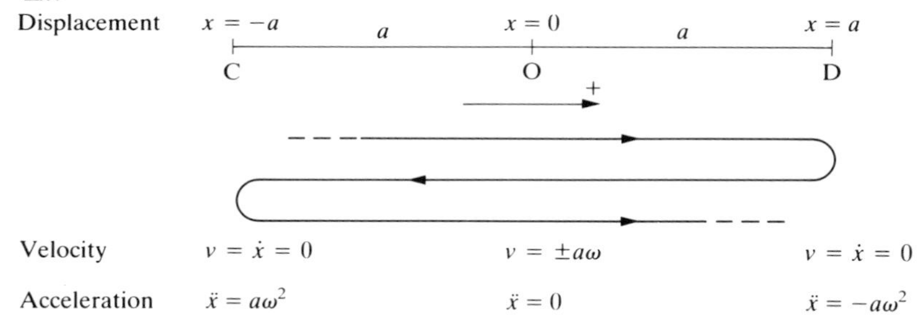

Figure 3.5

EXAMPLE

The lowest note on a double bass has a frequency of approximately 30 cycles per second. What is the maximum speed of a point on the string which is oscillating with SHM at this frequency and with an amplitude 0.5 cm?

Solution

The frequency is $\dfrac{\omega}{2\pi} = 30$

So $\omega = 30 \times 2\pi$

The maximum speed is given by $a\omega = 0.005 \times 60\pi$
$= 0.94\,\text{m s}^{-1}$ (to 2 decimal places)

EXAMPLE

In a harbour the cycle of tides can be modelled as SHM with a period of 12 hours 30 minutes. On a certain day high water is 10 m above low water.

(i) Sketch a graph of the height x m of the water above (or below) the mean level against t, the time in hours since the water was at mean level and rising.

(ii) Find a suitable expression to model x in terms of t.

(iii) Determine for how long the water was more than 6 metres above the low water mark.

(iv) Find the rate at which the tide is rising or falling when the water is 6 m above the low water mark.

(v) Find the maximum rate at which the tide rises.

Solution

(i)

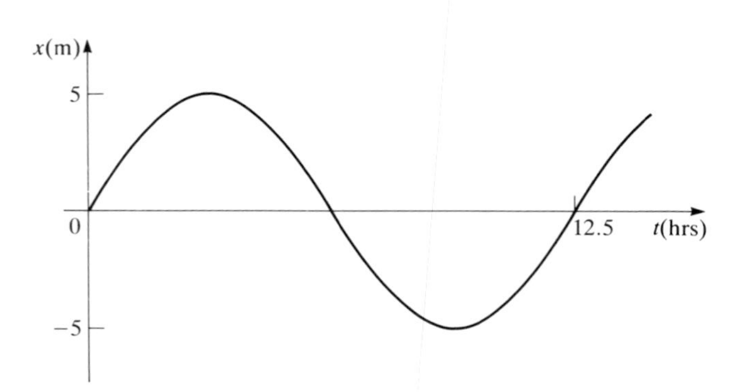

(ii) First find ω and a.

The period of the tide $T = 12.5$ hours.

Since $$T = \frac{2\pi}{\omega},\ \omega = \frac{2\pi}{T}$$

Since T is in hours, ω is in radians per hour

$$\omega = 0.503$$

From the graph, the amplitude of the oscillation is 5 metres, so $a = 5$.

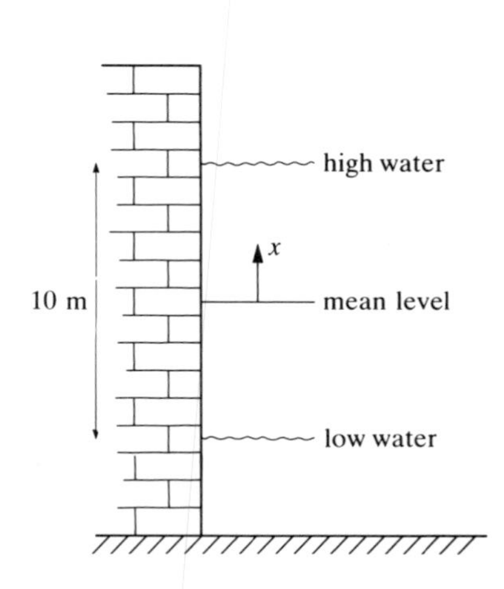

Since the water has risen x m in t hours after it was at the mean level, a suitable equation would be

$$x = a\sin \omega t$$
$$x = 5\sin(0.503t)$$

(iii) Remember that we are working with time in hours, so any values of t will be in hours. The tide is 6 m above low water when $x \geq 1$.

Now $x = 1$ when $\quad 5\sin(0.503t) = 1$

$$\Rightarrow \quad \sin(0.503t) = 0.2 = \sin 0.201$$

The first two values of t satisfying this equation are t_1 and t_2 where

$$0.503t_1 = 0.201 \quad \Rightarrow \quad t_1 = 0.400$$

and $\quad 0.503t_2 = \pi - 0.201 \quad \Rightarrow \quad t_2 = 5.846$

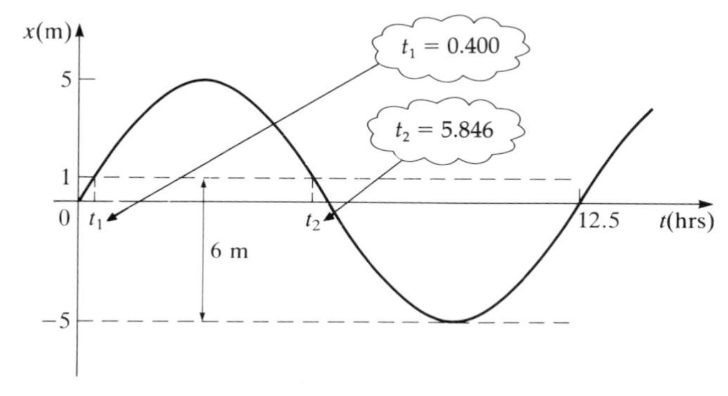

The water is 6 m above its lowest level for $(t_2 - t_1)$ hours i.e. about $5\frac{1}{2}$ hours.

(iv) The water is 6 m above its lowest level when $x = 1$. The velocity of the oscillation at this point is given by:

$$v^2 = \omega^2(a^2 - x^2)$$
$$\Rightarrow \quad v = \pm 0.503\sqrt{(25 - 1)}$$
$$= \pm 2.46$$

The rate of rise or fall is 2.46 m per hour, or about 4 cm per minute.

(v) Since $v^2 = \omega^2(a^2 - x^2)$, the maximum value of v is $a\omega$ when $x = 0$.

The water is rising fastest at the centre of its motion and the rate at which it is rising is:

$$a\omega = 0.503 \times 5$$
$$= 2.515 \text{ m per hour}$$
$$= 4.2 \text{ cm per minute.}$$

NOTE

A tidal range of 10 m is large, but is realistic for several places around the British coastline. The rate at which the water is rising, 4.2 cm per minute, does not sound high, but if you think about how fast the water would approach your deck chair up a typical beach (with a gradient of say 3°), you will see that it is quite dramatic.

For Discussion

The wave forms of two notes of music, A and B, are shown below.

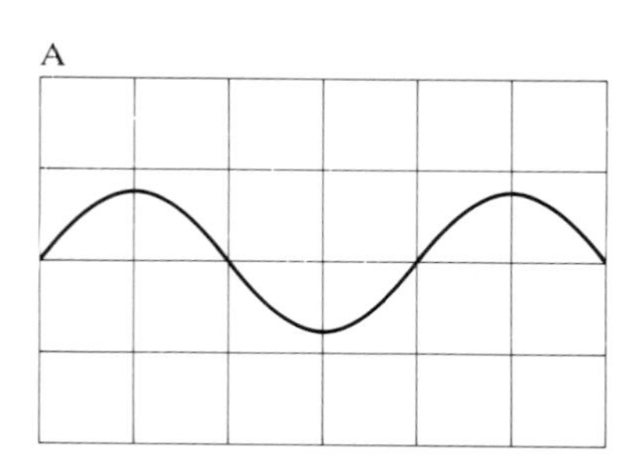

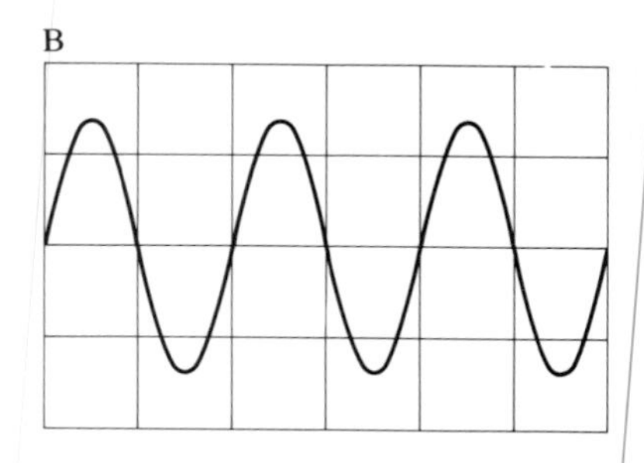

Which of the following statements are true?
(a) A is higher than B.
(b) A is louder than B.
(c) The period of A is twice as long as that of B.
(d) The amplitude of B is larger than that of A.
(e) The frequency of A is twice that of B.

Exercise 3A

1. The diagram illustrates the simple harmonic motion of a particle.

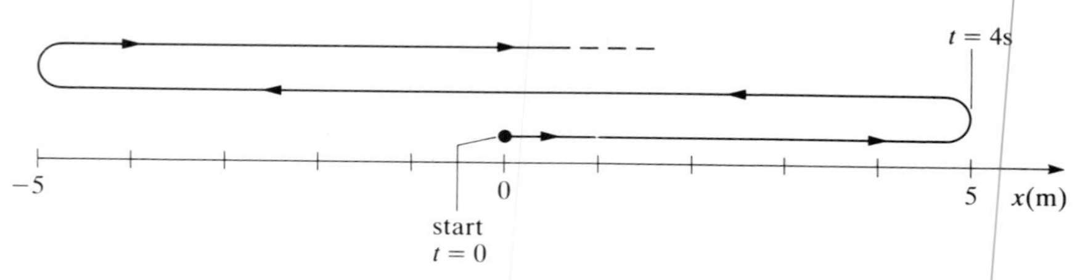

(i) Write down the amplitude of the simple harmonic motion.
(ii) Write down the period.
(iii) The motion can be described by the equation

$$x = a\sin \omega t$$

Write down the values of a and ω.
(iv) Find (a) the displacement (b) the velocity and (c) the acceleration of the particle when $t = 0$.

2. A particle is undergoing simple harmonic motion with period 12 seconds and amplitude 2 metres. The motion can be described by the equation $x = a\sin \omega t$.
(i) Write down the values of a and ω.
(ii) What distance does the particle travel in the first minute?
(iii) What is the maximum speed of the particle?
(iv) What is the displacement from the central point of the particle when it is travelling at half its maximum speed?

3. A particle is performing simple harmonic motion with 5 complete cycles per second. At a particular instant it is stationary. It travels 16 cm before it is next stationary.

(i) Write down the period and amplitude of the motion.

(ii) Its motion can be described by the equation

$$x = a\sin \omega t$$

Write down the values of a and ω and state the position and velocity of the particle when timing was started (i.e. at $t = 0$).

(iii) Does the speed of the particle ever exceed $250\,\text{cms}^{-1}$?

(iv) What is the displacement of the particle from its central position when its velocity is $4\,\text{cms}^{-1}$?

4. A particle is performing simple harmonic motion described by the equation

$$x = a\sin \omega t$$

Its maximum speed is $10\,\text{cms}^{-1}$ and its period is $\frac{\pi}{2}$ seconds.

(i) Find the values of a and ω.

(ii) Find the acceleration of the particle when its displacement from its central position is 2.5 cm.

(iii) Find its speed when its displacement is 2 cm.

(iv) Find how far the particle travels during a time of

(a) 1 second

(b) 2 seconds

after leaving the central position.

5. The motion of a particle is described by the equation

$$x = a\sin \omega t$$

(i) Find, by differentiation, an expression for its velocity, v, at time t.

(ii) Show that $v = \pm\omega\sqrt{(a^2 - x^2)}$ and explain the significance of the plus or minus sign.

(iii) Find, by further differentiation, an expression for the acceleration, a, of the particle at time t.

(iv) Show that $\ddot{x} = -\omega^2 x$ and comment on the significance of the minus sign.

6. (i) Sketch graphs of

(a) displacement (from the central position) against time,

(b) velocity against time and

(c) acceleration against time

for a particular simple harmonic motion. Line the three graphs up vertically and use the same scale for time in all three.

All three quantities (displacement, velocity and acceleration) have maximum positive, zero, and maximum negative values.

(ii) Look at your graphs and state which of these events occur at the same time.

7. A piston in an engine oscillates with a period of 0.03 seconds and an amplitude of 0.35 m. Modelling the oscillations as simple harmonic motion:

Exercise 3A continued

(i) draw a sketch graph to illustrate the oscillations;
(ii) calculate the frequency of the motion;
(iii) calculate the maximum speed of the piston.

8. One end of a metal strip is clamped while the free end vibrates in SHM with a frequency of $30\,\text{Hz}$ and amplitude $5\,\text{mm}$.
(i) Find the period of the motion.
(ii) Find the maximum speed of the vibrating end.
(iii) Find the maximum acceleration of the vibrating end.

9. Air sickness may be caused by the rhythmic vibrations of an aircraft. It has been observed that about 50% of the passengers of an aircraft suffer air sickness when it bounces up and down with a frequency of about $0.3\,\text{Hz}$ and a maximum acceleration of $4\,\text{ms}^{-2}$. Assuming that the motion is simple harmonic, find
(i) the period of the motion;
(ii) the amplitude of the motion;
(iii) the greatest vertical speed during this motion.

10. A jig-saw operates at 3000 strokes per minute with the tip of the blade moving $17\,\text{mm}$ from the top to the bottom of a stroke. (One stroke is a complete cycle.) Assuming that the motion is simple harmonic, find
(i) the maximum speed of the blade;
(ii) the maximum acceleration of the blade;
(iii) the speed of the blade when it is $6\,\text{mm}$ from the central position.

11. A loudspeaker cone sounding a pure note of frequency $2000\,\text{Hz}$ is modelled by simple harmonic motion of amplitude $2\,\text{mm}$.
(i) Calculate the period and angular frequency of the cone.
(ii) Calculate the maximum speed and maximum acceleration of the cone.

12. Musical notes which are an octave apart have frequencies in the ratio $1{:}2$. The note A above middle C has a frequency of $440\,\text{Hz}$. On a full-size keyboard there are 4 "A"s below it and 3 above it (the range on the keyboard is just over 7 octaves).
(i) Work out the values of ω corresponding to the frequencies of these seven "A"s.
(ii) Find the maximum speed of points on piano strings which are vibrating with amplitude $1\,\text{mm}$ to produce the highest and lowest of these notes.

Alternative forms of the equation for SHM

In the previous section you saw that $x = a\sin \omega t$ is a solution of the SHM differential equation $\ddot{x} = -\omega^2 x$. The activity on page 79 showed you that

$x = a\cos \omega t$ is also a solution. In this section we look at a number of alternative forms of the solution, in each case showing that they satisfy the SHM differential equation.

The forms $x = a\sin(\omega t + \varepsilon)$ and $x = a\cos(\omega t + \varepsilon)$

Since the graph of $x = a\sin \omega t$ passes through the origin, it represents SHM in which the particle is at the centre of its oscillation at time zero. Similarly, $x = a\cos \omega t$ represents SHM in which the particle is initially at maximum positive displacement. There will be times when you need to write an expression for SHM which starts at some other point in the cycle, and this is done by introducing a phase shift of an angle ε (the Greek letter 'epsilon'). This is equivalent to a phase shift of time $\frac{\varepsilon}{\omega}$.

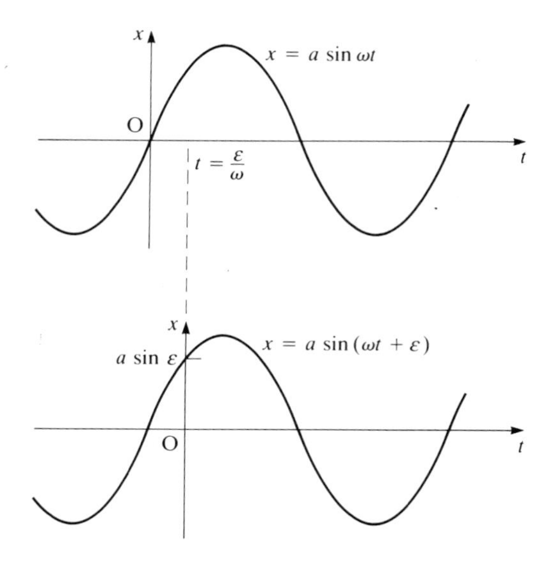

Figure 3.6

The equation $x = a\sin \omega t$ then becomes $x = a\sin(\omega t + \varepsilon)$, and the effect of this on the graph is shown in figure 3.6. The effect on $x = a\cos \omega t$ is similar.

EXAMPLE

The functions f and g are given by

$$\text{f: } x = a\sin(\omega t + \varepsilon) \text{ and g: } x = a\cos(\omega t + \varepsilon)$$

where ε (epsilon) is a positive constant.

For each of these functions:

(i) Differentiate it with respect to time to find v (or $\dot{x}$) and show that $v^2 = a^2(\omega^2 - x^2)$.

(ii) Differentiate v with respect to time to find $\ddot{x}$ and show that $\ddot{x} = -\omega^2 x$.

Solution

(i) For f:

$$x = a\sin(\omega t + \varepsilon)$$
$$\Rightarrow \quad v = \dot{x} = a\omega\cos(\omega t + \varepsilon) \qquad ①$$

Therefore
$$v^2 = a^2\omega^2\cos^2(\omega t + \varepsilon)$$
$$= \omega^2[a^2 - a^2\sin^2(\omega t + \varepsilon)]$$
$$\Rightarrow \quad v^2 = \omega^2(a^2 - x^2)$$

Similarly for g:

$$x = a\cos(\omega t + \varepsilon)$$
$$\Rightarrow \quad v = \dot{x} = -a\omega\sin(\omega t + \varepsilon) \qquad ②$$

Therefore
$$v^2 = a^2\omega^2\sin^2(\omega t + \varepsilon)$$
$$= \omega^2[a^2 - a^2\cos^2(\omega t + \varepsilon)]$$
$$\Rightarrow \quad v^2 = \omega^2(a^2 - x^2)$$

(ii) Differentiating equation ① gives

$$\ddot{x} = -a\omega^2\sin(\omega t + \varepsilon)$$
$$\Rightarrow \quad \ddot{x} = -\omega^2 x$$

Similarly differentiating ② gives:

$$\ddot{x} = -a\omega^2\cos(\omega t + \varepsilon)$$
$$\Rightarrow \quad \ddot{x} = -\omega^2 x$$

These results show that the functions f and g represent SHM

The form $x = A\sin\omega t + B\cos\omega t$

The function $x = A\sin \omega t + B\cos \omega t$ represents the sum of two simple harmonic motions, $A\sin\omega t$ and $B\cos\omega t$. These have the same period $\left(\frac{2\pi}{\omega}\right)$ and frequency, but different amplitudes (A and B). They are a quarter of a cycle out of phase.

What is the effect of adding two SHMs in this way?

By differentiating twice with respect to time, you can show that the function obeys the SHM equation $\ddot{x} = -\omega^2 x$.

$$x = A\sin \omega t + B\cos \omega t$$
$$\dot{x} = A\omega\cos \omega t - B\omega\sin \omega t$$
$$\ddot{x} = -A\omega^2\sin \omega t - B\omega^2\cos \omega t$$

This expression for $\ddot{x}$ may be written as

$$\ddot{x} = -\omega^2(A\sin \omega t + B\cos \omega t)$$

and so
$$\ddot{x} = -\omega^2 x.$$

This proves that $x = A\sin \omega t + B\cos \omega t$ represents SHM with period $\frac{2\pi}{\omega}$. To show that this form is equivalent to $x = a\sin(\omega t + \varepsilon)$, use the following technique (which you may already have met in *Pure Mathematics 3*, chapter 2).

The function $x = A\sin \omega t + B\cos \omega t$ is rewritten as

$$x = \sqrt{(A^2 + B^2)}\left(\sin \omega t \frac{A}{\sqrt{(A^2 + B^2)}} + \cos \omega t \frac{B}{\sqrt{(A^2 + B^2)}}\right)$$

In the right-angled triangle shown in figure 3.7, for the case when A and B are both positive $\cos \varepsilon = \frac{A}{\sqrt{(A^2 + B^2)}}$ and $\sin \varepsilon = \frac{B}{\sqrt{(A^2 + B^2)}}$. The hypotenuse is $\sqrt{(A^2 + B^2)} = a$.

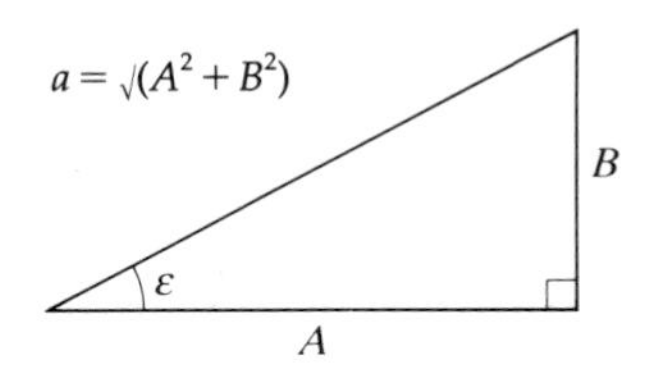

Figure 3.7

Using these results gives

$$x = a(\sin \omega t \cos \varepsilon + \cos \omega t \sin \varepsilon).$$

Using the compound angle formula $\sin(\theta + \phi) = \sin\theta\cos\phi + \cos\theta\sin\phi$, this can be written

$$x = a\sin(\omega t + \varepsilon).$$

Thus the SHM $x = A\sin \omega t + B\cos \omega t$ is equivalent to the SHM $x = a\sin(\omega t + \varepsilon)$, where the amplitude a is $\sqrt{(A^2 + B^2)}$, and the phase angle ε is given by $\sin \varepsilon = \frac{B}{\sqrt{(A^2 + B^2)}}$ and $\cos \varepsilon = \frac{A}{\sqrt{(A^2 + B^2)}}$.

This is also equivalent to a cosine form, since for any angle α

$$\sin \alpha = \cos\left(\alpha - \frac{\pi}{2}\right)$$

and so x can be written as

$$x = a\cos\left(\omega t + \varepsilon - \frac{\pi}{2}\right)$$

$$= a\cos(\omega t + \varepsilon') \text{ where } \varepsilon' = \varepsilon - \frac{\pi}{2}.$$

(If ε is an acute angle, ε' is negative.)

This means that the effect of adding together two SHMs of the same period is to create a single SHM, also with the same period but with greater amplitude. This is shown in figure 3.8 for $x = 3\cos t$ and $x = 4\sin t$.

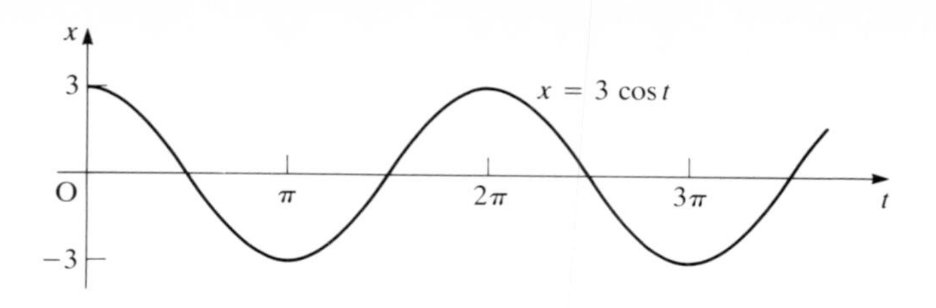

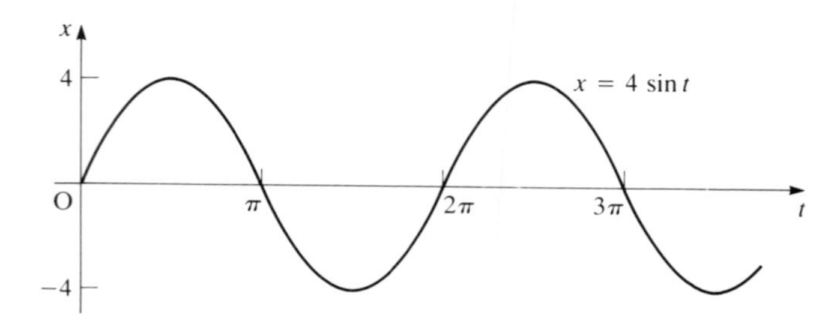

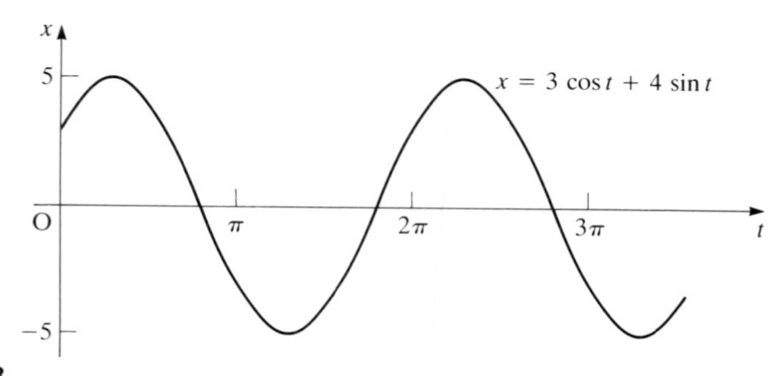

Figure 3.8

For Discussion

When you hear several musical instruments play notes of the same pitch, the vibrations which reach your ears are a combination of oscillations of different amplitudes. In addition, these may have been set in motion at different stages in the cycle and so be out of phase.

How is it possible for the music to sound pleasant?

The form $x = x_0 + a\sin \omega t$

Although the equation $x = x_0 + a\sin \omega t$ does not satisfy the SHM equation $\ddot{x} = -\omega^2 x$, it does represent SHM about the fixed point x_0 as you can see from figure 3.9.

You may find it helpful to think of this motion in terms of a new variable, z, representing the displacement from the central position. This is given by

$$\begin{aligned} z &= x - x_0 \\ &= x_0 + a\sin \omega t - x_0 \\ &= a\sin \omega t. \end{aligned}$$

The variable z does satisfy the SHM equation $\ddot{z} = -\omega^2 z$ and all the standard SHM results also hold for z.

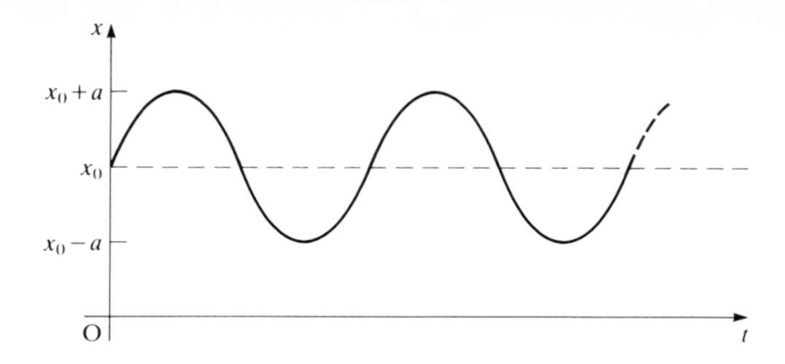

Figure 3.9

Choosing the most appropriate function to model a particular oscillation

You have seen that any particular example of SHM can be described using at least one of a variety of functions.

SHM about $x = 0$ with period $\frac{2\pi}{\omega}$, amplitude a	**SHM about $x = x_0$ with period $\frac{2\pi}{\omega}$, amplitude a**
$x = a\sin(\omega t + \varepsilon)$ $x = a\cos(\omega t + \varepsilon)$ $x = A\sin \omega t + B\cos \omega t$ where $\sqrt{(A^2 + B^2)} = A$	$x = x_0 + a\sin(\omega t + \varepsilon)$ $x = x_0 + a\cos(\omega t + \varepsilon)$ $x = x_0 + A\sin \omega t + B\cos \omega t$ where $\sqrt{(A^2 + B^2)} = a$

The constants a, A, B and ε are determined by the *initial conditions*, that is the speed, direction and displacement at time zero.

You will find it helpful to sketch the graph of the oscillation. This will show you the initial conditions and help you to choose the most appropriate form.

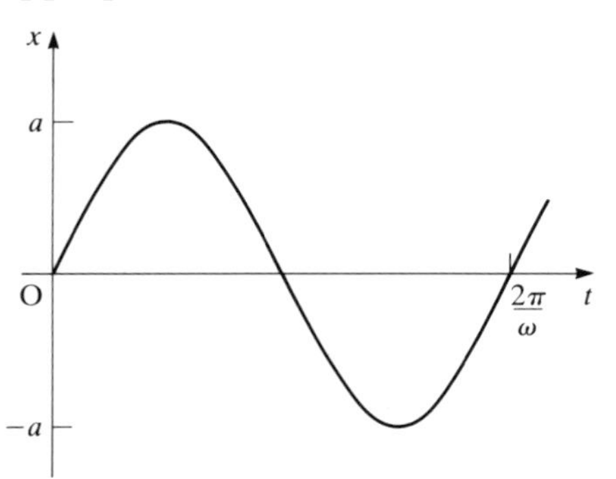

Figure 3.10

In this case when $t = 0$, $x = 0$ and $\dot{x} > 0$: the most appropriate form is

$$x = a\sin \omega t$$

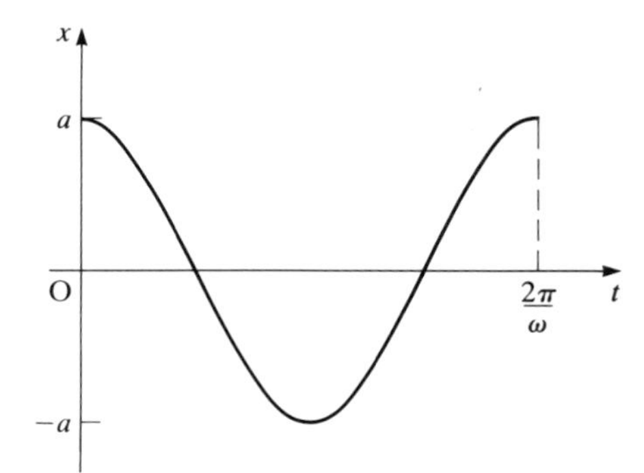

Figure 3.11

In this case when $t = 0$, $x = a$ and $\dot{x} = 0$: the most appropriate form is

$$x = a\cos \omega t$$

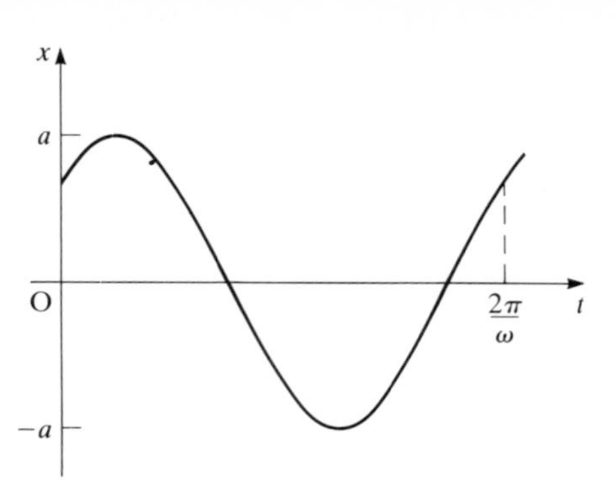

Figure 3.12

When the oscillation starts somewhere between the centre and an extreme, i.e. neither $x = 0$ nor $\dot{x} = 0$ when $t = 0$, the most appropriate form will be either

$$x = a\sin(\omega t + \varepsilon)$$

or $$x = a\cos(\omega t + \varepsilon)$$

or $$x = A\sin\omega t + B\cos\omega t$$

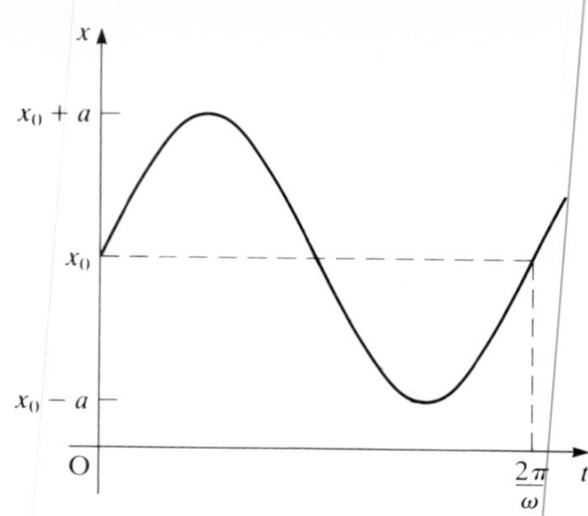

Figure 3.13

When the centre of the oscillations is not at the origin but at a point x_0, the appropriate equation will be one of those above but with x_0 added on. In this case

$$x = x_0 + a\sin\omega t$$

For Discussion

The graphs below show cases where the initial conditions are different from those covered above. What are the most appropriate forms to model these oscillations?

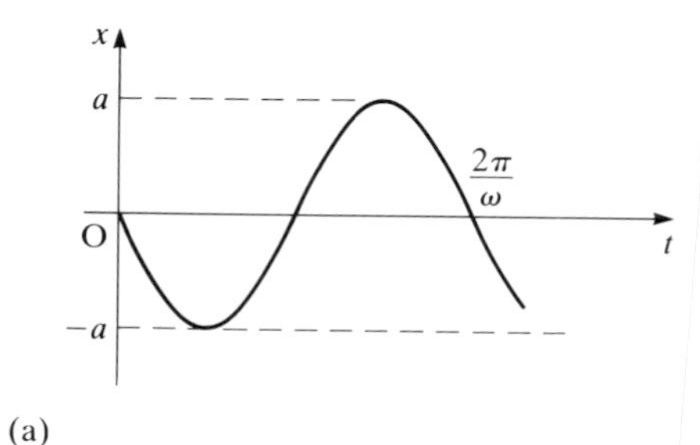

(a)

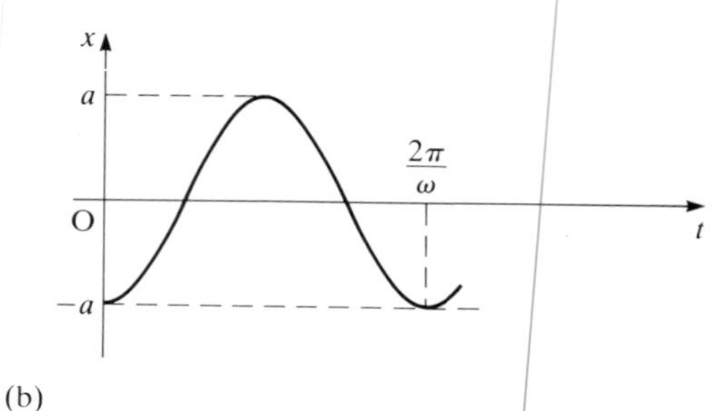

(b)

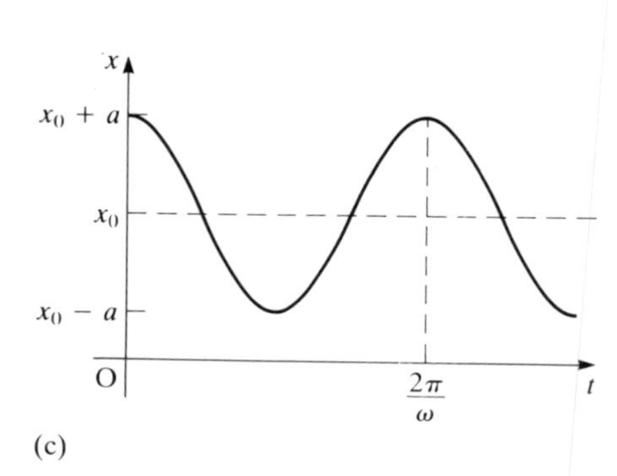

(c)

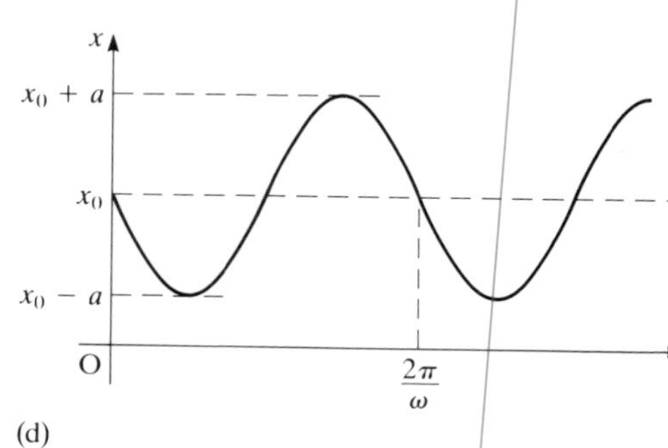

(d)

EXAMPLE

A particle is moving with SHM of period π. Initially it is $10\,\text{cm}$ from the centre of the motion and moving in the positive direction with a speed of $6\,\text{cm s}^{-1}$. Find an equation to describe the motion.

Solution

The information given is shown in the diagram below.

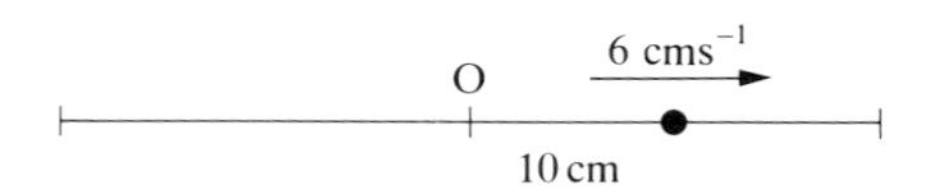

The initial speed is positive, so an appropriate equation is

$$x = a\sin(\omega t + \varepsilon),$$

and we need to find the values of a, ω, and ε.

Finding ω

Since the period of the motion is π,

$$\frac{2\pi}{\omega} = \pi \quad \Rightarrow \quad \omega = 2.$$

Finding a

Using
$$v^2 = \omega^2(a^2 - x^2)$$
$$6^2 = 2^2(a^2 - 10^2)$$
$$\Rightarrow \quad a = \sqrt{109}$$

Finding ε

Substituting $t = 0$ in $x = a\sin(\omega t + \varepsilon)$ gives

$$10 = \sqrt{109}\sin\varepsilon$$
$$\Rightarrow \quad \varepsilon = 1.28 \text{ radians (see Note below).}$$

So the equation for the motion is
$x = \sqrt{109}\sin(2t + 1.28)$.

NOTES

When finding ε you must be careful that you have selected the correct root of the equation. In this case at $t = 0$ the particle has positive displacement and positive velocity (it is on its way out and not on its way back), so $t = 0$ corresponds to an angle between 0 and $\frac{\pi}{2}$. The next root of the equation $10 = \sqrt{109}\sin\varepsilon$ is $(\pi - 1.28)$. This lies between $\frac{\pi}{2}$ and π and would be the correct value if the particle were on its way back, with displacement $+10$ and velocity -6.

SHM as the projection of circular motion

There is a close relationship between circular motion at constant speed and SHM. This can be illustrated by rotating a bob on the end of a string in a horizontal circle with constant angular velocity, thus forming a conical pendulum. If this is done between the light of an overhead projector and the wall, the shadow of the bob on the wall will perform SHM. (For true SHM the rays of light should be parallel and an approximation to this can be achieved if the pendulum is close to the wall and the overhead projector is as far away as possible.)

Assuming the rays of light are parallel, figure 3.14 shows the position, C, of the bob and its shadow, P, at a particular instant. As the bob moves round the circle from A to B, the shadow moves along the straight line from L to N.

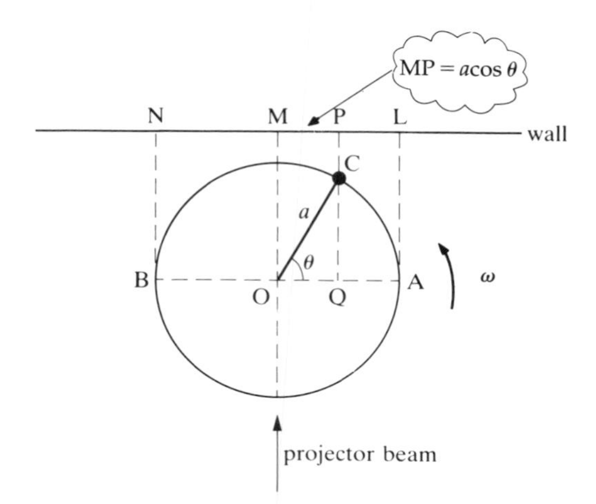

Figure 3.14

Assuming that $t = 0$ when the bob is at A, the angle θ is given by ωt, where ω is the angular speed of the bob. Thus

$$\text{MP} = a\cos\omega t.$$

This is one of the standard forms of SHM.

NOTE

The projection of uniform circular motion onto a straight line, illustrated here, is the only case where ω actually does represent a physical angular velocity.

EXAMPLE

An astronomer observes a faint object close to a star. Continued observations show the object apparently moving in a straight line through the star as shown in the diagram

star

faint object

The astronomer is able to estimate the apparent distance of the object from the centre of the star and records this at 30 day intervals, resulting in the following table:

Day:	t	0	30	60	90	120
Distance (10^{11} m):	x	1.0	1.5	1.9	2.0	1.9

The astronomer thinks that the object is a planet moving around the star in a circular orbit and that she is observing it from a point in the plane of the orbit. She decides to model the apparent distance from the centre by the SHM equation:

$$x = 2\sin(\omega t + \varepsilon)$$

(i) Use the values of x for $t = 0$ and $t = 90$ to find values for the constants ω and ε and verify that the other values of x are consistent with this model.

(ii) Assuming the model is correct, find
  (a) the radius of the orbit
  (b) the speed of the planet
  (c) the number of earth days the planet takes to go round its star.

[MEI]

Solution

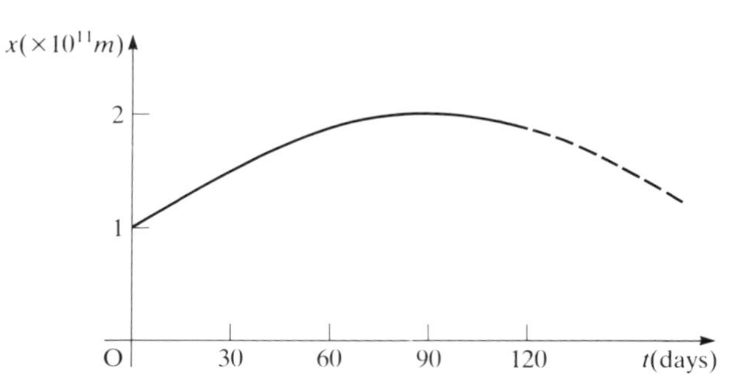

(i) Using $x = 2\sin(\omega t + \varepsilon)$:
When $t = 0$, $x = 1$ $\Rightarrow$ $1 = 2\sin\varepsilon$

$\Rightarrow$ $\varepsilon = \dfrac{\pi}{6}$

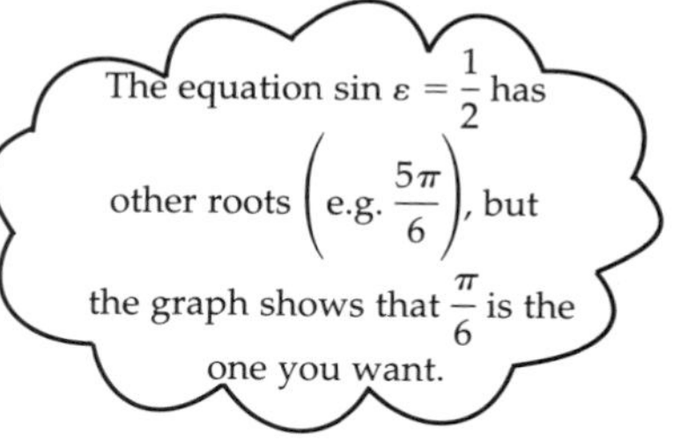

When $t=90$, $x=2$ $\Rightarrow$ $2=2\sin(90\omega+\frac{\pi}{6})$

$$\Rightarrow \sin(90\omega+\frac{\pi}{6})=1$$

$$\Rightarrow 90\omega+\frac{\pi}{6}=\frac{\pi}{2}$$

$$\Rightarrow \omega=\frac{\pi}{270}$$

So the model is $x=2\sin\left(\frac{\pi t}{270}+\frac{\pi}{6}\right)$

For the other values of t:

t	**observed displacement**	**model's prediction**
30	1.5	$2\sin\left(\frac{\pi}{9}+\frac{\pi}{6}\right)=1.53$
60	1.9	$2\sin\left(\frac{2\pi}{9}+\frac{\pi}{6}\right)=1.88$
120	1.9	$2\sin\left(\frac{4\pi}{9}+\frac{\pi}{6}\right)=1.88$

This shows that the model is a very good predictor of the actual position of the object.

(ii) (a) The radius of the orbit is the amplitude a of the motion:

$$\text{radius}=2\times10^{11}\,\text{m}$$

(b) The linear speed is

$$a\omega=2\times10^{11}\times\frac{\pi}{270}=2.3\times10^{9}\,\text{ms}^{-1}$$

(c) The total time for one orbit is the period of the SHM i.e. $\frac{2\pi}{\omega}$

$$\text{Number of days}=2\pi\div\frac{\pi}{270}$$

$$=540\text{ days}$$

HISTORICAL NOTE

In 1822, the French mathematician J B Fourier (1768–1830) showed that any function of t can be written as a sum of sines of multiples of t and this is now called a Fourier series. It follows that any vibration which can be written as a function of t can be reproduced by adding simple harmonic vibrations. Fourier accompanied Napoleon to Egypt in 1798 and was made a baron ten years later. He discovered this theorem while working on the flow of heat.

Exercise 3B

1. (i) Sketch graphs of the following functions:
   (a) $x = 3\cos t$
   (b) $x = 3\cos 2t$
   (c) $x = 3\cos t + 4\sin t$
   (d) $x = 5\cos(0.2t + \frac{\pi}{4})$
   (ii) In each case write down the amplitude, period and frequency of the oscillations.

2. Find a suitable simple harmonic model giving x(in metres) in terms of t(in seconds) for each of the following.
   (i) Period = 3 s, and initially $\dot{x} = 0$ and $x = 5$.
   (ii) Period = 2 s, amplitude = 6 m, and initially $x = 0$.
   (iii) Period = 5 s, amplitude = 4 m, and initially $x = 2$.
   (iv) Period = 4 s, and initially $x = 1.5$ and $\dot{x} = 6$.

3. The data in the following table is taken from a stroboscopic photograph of an oscillating object on an elastic spring. The displacement of the object from an equilibrium level was measured at intervals of 0.1 s.

t s	0.1	0.2	0.3	0.4	0.5	0.6	0.7	0.8	0.9	1.0	1.1	1.2	1.3
x mm	49	22	−14	−44	−59	−54	−29	5	38	58	57	37	3

   (i) Plot the displacement of the object against time on graph paper.
   (ii) Estimate the amplitude, period and frequency of the motion.

4. The displacements of a particle along a straight line at different times are given in the table below.

t	0	1	2	3	4	5	6	7	8	9
x	3.5	5	3.5	0	−3.5	−5	−3.5	0	3.5	5

Exercise 3B continued

(i) These data may be modelled using the SHM equation

$$x = a\cos(\omega t - \varepsilon).$$

Find the values of a, ω and ε.

(ii) An alternative equation is $x = a\sin(\omega t + \delta)$

Find the value of δ.

(iii) A third equation to describe the data is

$$x = b\sin\omega t + b\cos\omega t$$

Find the value of b and show that this form of the equation does indeed describe the data.

5. A particle is executing simple harmonic motion. Its position, y (in metres), at time t (in seconds) is given in the table below.

t	2	4	6	8	10	12	14	16
y	8	5.4	5.4	8	10.6	10.6	8	5.4

(i) Draw the graph of y against t and write down the central value of y for the motion.

The motion may be described by the equation

$$y = y_0 + a\cos(\omega t + \varepsilon)$$

(ii) Use the information in the table to determine the values of a, ω, and ε.

6. A simple harmonic motion is described by the equation

$$x = 2\cos\frac{\pi}{4}t + \sqrt{5}\sin\frac{\pi}{4}t$$

(i) Plot the graph of x against t for $0 \le t \le 20$.

The motion may be described by an equation of the form

$$x = a\sin(\omega_1 t + \varepsilon)$$

(ii) Determine the values of a, ω_1 and ε.

Another way of describing the motion is given by the equation

$$x = b\cos(\omega_2 t + \delta)$$

(iii) State which of b, ω_2 and δ are equal, respectively, to a, ω_1 and ε.

(iv) State the values of b, ω_2 and δ.

7. As a result of storms in different places, two swell wave patterns, both running in the same direction, occur at the same time over a stretch of open sea. Their heights above the mean sea level can be modelled as follows.

Wave pattern A: $h_A = 1.5\sin\frac{\pi}{15}t$ (t is in seconds, h is in metres)

Wave pattern B: $h_B = 2\cos\frac{\pi}{15}t$

(i) Plot the two wave patterns on the same piece of graph paper taking values of t from 0 to 45 at 5 second intervals.

(ii) The overall height of the water, h, is given by

$$h = h_A + h_B$$

Plot the values of h for $0 \leq t \leq 45$ on the same graph that you used for h_A and h_B.

(iii) Show algebraically that the effect of the two wave patterns is that of a single wave pattern described by

$$h = a\sin(\omega t + \varepsilon)$$

and state the values of a, ω and ε.

8. The displacement of a particle along a straight line at time t is given by

$$x = 4\cos\left(\frac{\pi t}{12} + \frac{\pi}{6}\right)$$

(i) Sketch the displacement–time curve.

(ii) An alternative description is given by

$$x = a\sin(\omega t + \varepsilon)$$

Write down the values of a, ω and ε.

(iii) Is it possible to describe x by the equation

$$x = p\sin\omega t + q\cos\omega t?$$

If so, find the values of p and q.

9. Data on sunspots show that their number in any year, N, follows an 11-year cycle. An approximate model gives N as

$$N = N_0 + a\cos\omega T$$

where T is the number of years (0, 1, 2, 3, . . .) since the year in which there was last a maximum. A maximum of 150 occurred in 1948 and the subsequent minimum was 20.

(i) Determine the appropriate values of N_0, a and ω from the given information.

(ii) Estimate the numbers of sunspots in 1950 and 1955.

10. The height, h metres, of water above the bottom of a harbour varies with the tide.

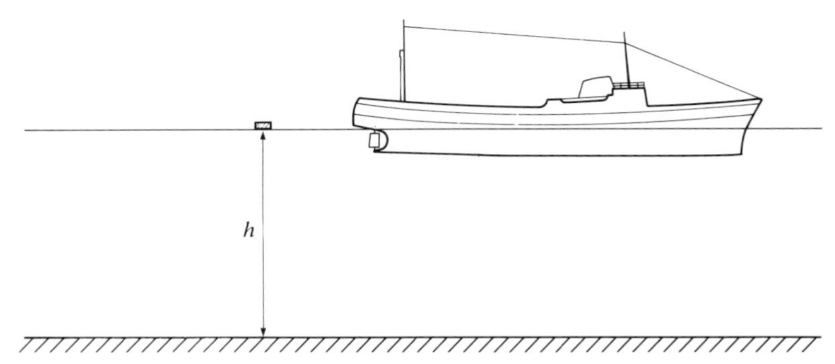

Exercise 3B continued

The vertical motion of a piece of wood floating on the surface may be modelled by the equation

$$h = h_0 + a\sin(\omega t + \varepsilon)$$

where t is the time in hours since midnight.

(i) Describe the vertical motion of the piece of wood in terms of h_0, a and ω.

The heights at certain times on one day are given by:

t (hours)	2	5	8	11	12
h (metres)	10	6	2	6	8

(ii) Find the values of h_0, a, ω and ε.
(iii) What are the vertical speed and acceleration of the piece of wood when $t = 8.6$?
(iv) What is the greatest speed at which the water level rises?

11. The motion of the fore and hind wings of a locust can be modelled approximately using the ideas of oscillation. The motion of the fore wings is modelled by

$$\alpha = 1.5 + 0.5\sin(1.05t - 0.005)$$

where α is the angle between the fore wing and the vertical, and t is measured in hundredths of a second.

(i) Find the period and the amplitude of the motion.

The hind wings initially make an angle of 1.5 radians to the vertical, then they oscillate with period 0.06 s and amplitude 1.5 radians.

(ii) Construct a model of the form

$$\beta = \beta_0 + a\sin(kt)$$

for the motion of the hind wings.

Oscillating mechanical systems

There are very many mechanical systems which can be modelled using SHM. Two of these are the spring–mass oscillator and the simple pendulum. The motion of the simple pendulum approximates to SHM for small angles as you will see in the next section.

The simple pendulum

This consists of a bob suspended on the end of a light inelastic string as illustrated by the apparatus in figure 3.15.

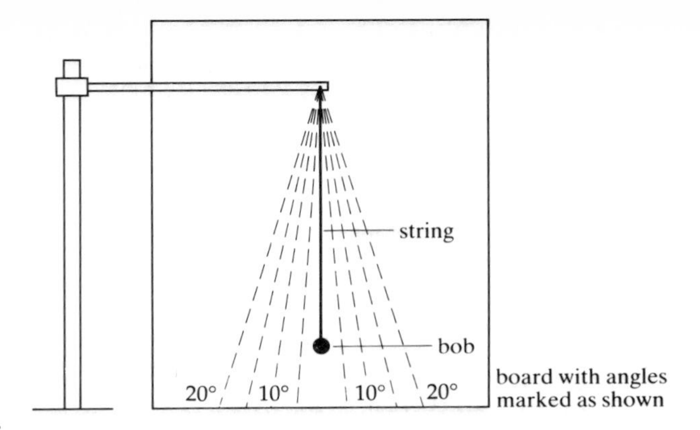

Figure 3.15

Experiment

Set up the apparatus as shown, making sure that the board is as close as possible to the pendulum without touching it. Use the apparatus to investigate whether the period of small oscillations depends on:

(i) the mass of the bob;
(ii) the length of the string;
(iii) the amplitude of the swing;
(iv) the starting point in the cycle.

In order to construct a mathematical model for the motion of a simple pendulum, take a general position some time t after it has been set in motion, when the string makes an angle θ with the vertical.

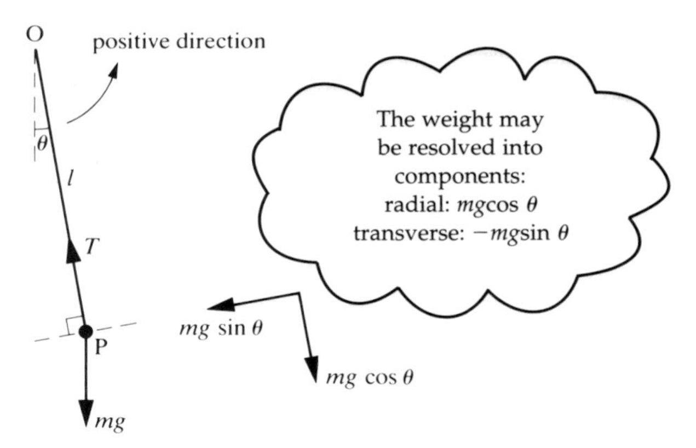

Figure 3.16

The forces acting on the bob are the tension in the string and the force of gravity mg, where m is the mass of the bob as shown in figure 3.16. It swings through a small arc of a circle of radius l where l is the length of the string.

There is no motion in the radial direction. In the transverse direction, the acceleration, $l\ddot{\theta}$, is given by

$$-mg\sin\theta = ml\ddot{\theta}$$

$$\Rightarrow \quad \ddot{\theta} = -\frac{g}{l}\sin\theta$$

When the angle is measured in radians, $\theta \simeq \sin\theta$ for small angles (up to about 0.3 radians for accuracy correct to 2 decimal places). In this case:

$$\ddot{\theta} = -\frac{g}{l}\theta$$

This is the standard equation for simple harmonic motion, $\ddot{x} = -\omega^2 x$, with x replaced by θ and ω^2 replaced by $\frac{g}{l}$.

A pendulum is usually set in motion by pulling the bob to one side, say to an angle α, and then releasing it from rest. If this is the case, $\theta = \alpha$ and $\dot{\theta} = 0$ when $t = 0$.

The appropriate form of the SHM equation is

$$\theta = \alpha\cos\sqrt{\frac{g}{l}}t.$$

The period is given by

$$T = \frac{2\pi}{\omega} = 2\pi\sqrt{\frac{l}{g}}$$

There are several points to note:

- The SHM involves the angular displacement, θ, rather than the linear displacement of the bob.
- SHM is a good model for small values of θ, but the approximation becomes less good with increasing θ.
- The mass of the bob is not present in the SHM equation, so its value does not affect the motion.
- The amplitude is specified only by the initial conditions.
- The period $2\pi\sqrt{\frac{l}{g}}$ is not affected by the amplitude of the motion or the mass of the bob. It depends only on g and the length of the string.
- It is important not to confuse the angular velocity, $\dot{\theta}$, of the pendulum with ω, the angular frequency of the SHM.

EXAMPLE

A simple pendulum consists of a mass hanging on the end of a light inextensible string. The angle θ radians between the string and the vertical satisfies the differential equation $\ddot{\theta} = -\frac{g}{l}\theta$.

(i) Find the length of the pendulum for which the period for small oscillations is 2 seconds.

The pendulum is released when the string makes an angle of 0.2 radians with the vertical.

(ii) Find an equation for θ in terms of t.

(iii) Find the times when the pendulum is in the equilibrium position.
(iv) Find also the time taken for the pendulum to move from the equilibrium position to a point half way to the end of the oscillation.

Solution

(i) The equation of motion, $\ddot{\theta} = -\frac{g}{l}\theta$, is the SHM differential equation

with $\omega^2 = \frac{g}{l}$. All the standard results for SHM apply.

The period is 2 s, so $\frac{2\pi}{\omega} = 2 \quad \Rightarrow \quad \omega = \pi.$

Therefore $$\pi^2 = \frac{g}{l}$$

$$\Rightarrow \quad l = \frac{g}{\pi^2} = \frac{9.8}{(3.14\ldots)^2}$$

The length of the pendulum is 0.993 m or about 1 metre.

(ii) The pendulum is released when $\theta = 0.2$ ($\theta = \alpha$ when $t = 0$).

The most appropriate function to use to model the motion is

$$\theta = 0.2\cos\omega t$$

We have already seen that

$$\omega = \pi$$

Therefore $$\theta = 0.2\cos\pi t$$

(iii) The pendulum is in the equilibrium position when $\theta = 0$, i.e. when

$$0.2\cos\pi t = 0$$
$$\Rightarrow \quad \cos\pi t = 0$$
$$\Rightarrow \quad \pi t = \frac{\pi}{2}, \frac{3\pi}{2}, \frac{5\pi}{2} \ldots$$
$$\Rightarrow \quad t = \frac{1}{2}, \frac{3}{2}, \frac{5}{2}, \ldots$$

The graph illustrates the first two swings of the pendulum.

It is first at the equilibrium position, $\theta = 0$, when $t = 0.5$. It is then half-way to the end of the oscillation when $\theta = -0.1$.

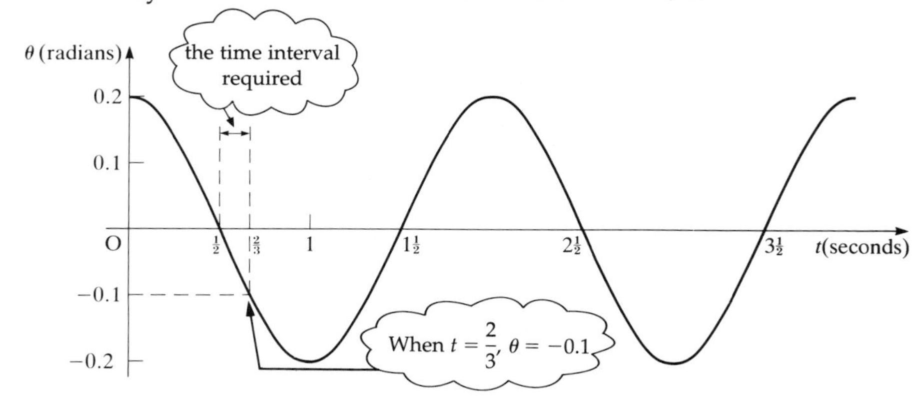

The value of t when $\theta = -0.1$ is given by

$$0.2\cos \pi t = -0.1$$

$$\Rightarrow \quad \pi t = \frac{2\pi}{3}$$

$$\Rightarrow \quad t = \frac{2}{3}.$$

The time taken from the centre to the half-way point is $\frac{2}{3} - \frac{1}{2} = \frac{1}{6}$ seconds.

NOTE

Notice that this is one third of and not half of the time taken to travel from the centre to the end. This is because, for SHM, the velocity is greater near the centre of the motion than near the ends.

The spring–mass oscillator

On page 75 you saw that the motion of a particle of mass m hanging from a spring of natural length l_0, modulus of elasticity, λ, is governed by the equation

$$\ddot{x} = -\frac{\lambda}{ml_0}x,$$

where x is the displacement from the equilibrium position, as in figure 3.17.

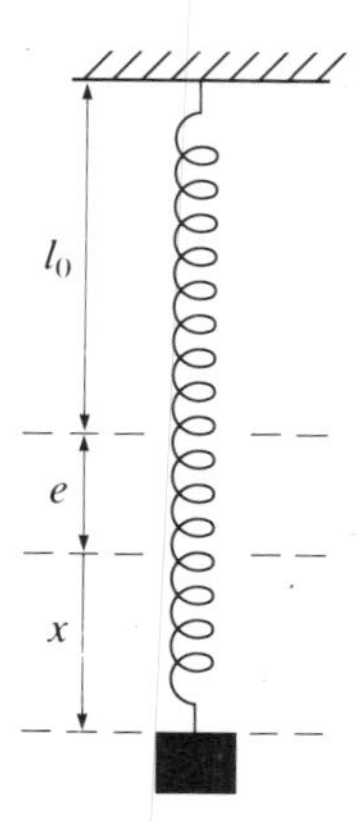

Figure 3.17

This is the standard equation for SHM with

$$\omega^2 = \frac{\lambda}{ml_0}$$

Its period is $\frac{2\pi}{\omega} = 2\pi\sqrt{\frac{ml_0}{\lambda}}$.

If k is the stiffness of the spring, $k = \frac{\lambda}{l_0}$ so this becomes $2\pi\sqrt{\frac{m}{k}}$.

This means that the period of the oscillations depends only on the mass and the stiffness of the spring.

Note that this model is good only so long as Hooke's law holds, but provided this is so, the motion is an example of true SHM, unlike that of the simple pendulum which only approximates to SHM.

EXAMPLE

A particle of mass 200 grams is attached to a light spring of natural length 40 cm and stiffness 50 Nm^{-1}. The particle is allowed to hang vertically in equilibrium.

(i) Find the extension of the spring in this position.

The spring is now pulled down 3 cm and released from rest.

(ii) Find the length of the spring as a function of time.

Solution

(i) The diagram shows the relevant lengths, and the forces acting on the particle in equilibrium. The extension in the spring is x_0 m.

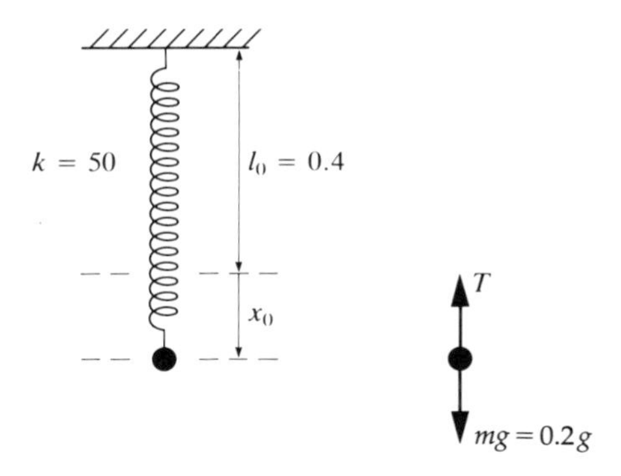

By Hooke's law: $T = kx_0 = 50x_0$

Since the particle is in equilibrium,

$$T = 0.2g$$

$$\Rightarrow \quad 50x_0 = 0.2 \times 9.8$$

$$x_0 = 0.0392$$

The extension is 0.0392 m, or 3.92 cm.

(ii) Let x m be the displacement from the equilibrium position in the downwards direction.

The extension of the spring is then $\quad x + x_0$

we know that the value of x_0 is 0.0392

By Hooke's law, $T = 50(x + x_0)$

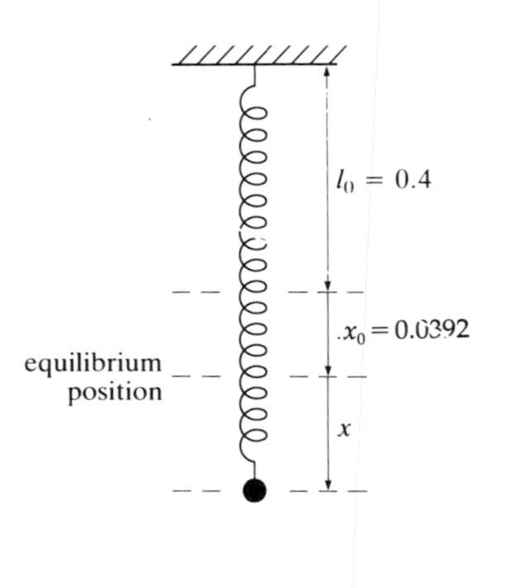

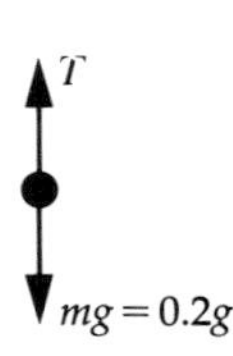

Applying Newton's 2nd law in the downwards direction,

$$0.2g - 50(x + x_0) = 0.2\ddot{x}$$

$$\Rightarrow \quad 0.2g - 50x - 50x_0 = 0.2\ddot{x}$$

This shows the advantage of measuring x from the equilibrium position. The weight and kx_0 cancel each other out

However, as we found in part (i), $0.2g = 50x_0$

and so the equation may be simplified to

$$0.2\ddot{x} = -50x$$

$$\Rightarrow \quad \ddot{x} = -250x.$$

This is the SHM equation with $\omega^2 = 250$, and so $\omega = \sqrt{250}$.

Since the particle starts at the end of the oscillation with $x = 0.03$, the appropriate equation is of the form $x = a\cos \omega t$, with $a = 0.03$ and $\omega = \sqrt{250}$.

So, at time t, $\quad x = 0.03\cos \sqrt{250}t$

and the total length is $0.4 + 0.0392 + 0.03\cos \sqrt{250}t$

$$= 0.4392 + 0.03\cos \sqrt{250}t.$$

For Discussion

Would the period of the oscillations in the above example be the same on the moon?

EXAMPLE

Two springs have the same modulus of elasticity of 30 N, but are of natural lengths 0.4 m and 0.6 m. An object of mass 0.5 kg is attached to one end of each spring and the other ends are attached to two points which are 1.2 m apart on a smooth horizontal table. Find the period of small oscillations of this system.

Solution

The vertical forces acting on the object, its weight and the normal reaction of the table, have no effect on the motion: they balance each other.

The first step is to find the equilibrium position of the object. The diagram shows the relevant lengths and the horizontal forces acting when the object is in equilibrium. The extensions in the springs are e_1 m and e_2 m and the tension, T newtons, is the same on both sides.

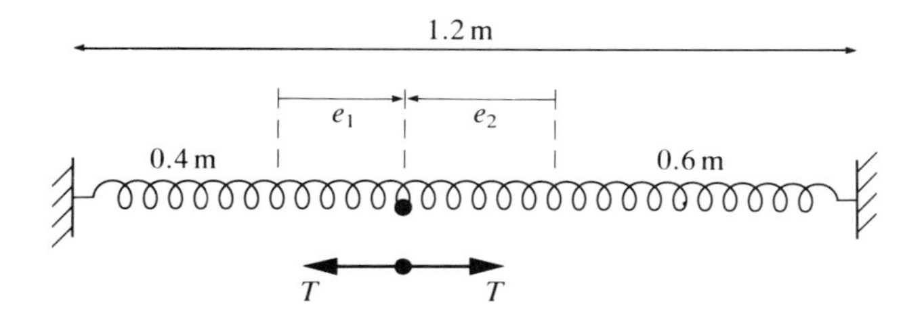

Hooke's law applied to each spring gives

$$T = \frac{30}{0.4} \times e_1 = 75e_1$$

and

$$T = \frac{30}{0.6} \times e_2 = 50e_2$$

$$\Rightarrow \quad 75e_1 = 50e_2$$

$$\Rightarrow \quad e_2 = 1.5e_1$$

The total length is 1.2 m, so $e_1 + e_2 = 0.2$,

$$\Rightarrow \quad 2.5e_1 = 0.2$$

$$\Rightarrow \quad e_1 = 0.08 \text{ and } e_2 = 0.12$$

The next step is to find the equation of motion, so it is necessary to consider a general position for the object, and this is given by the displacement x m from the equilibrium position. The direction towards the right is taken to be positive.

The tensions in the springs are now different.

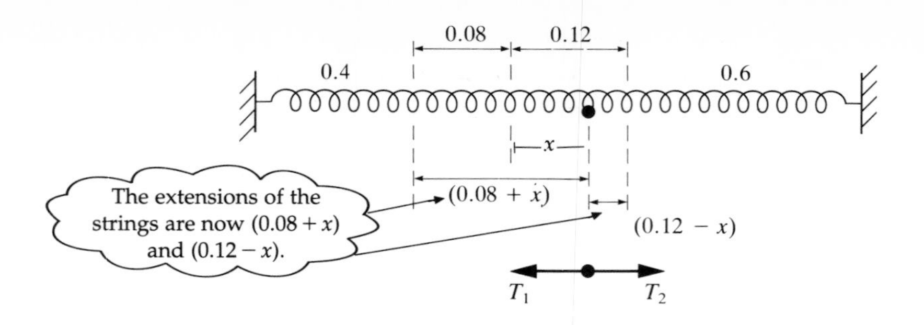

Hooke's law gives

$$T_1 = \frac{30}{0.4}(0.08 + x) = 75(0.08 + x)$$

$$T_2 = \frac{30}{0.6}(0.12 - x) = 50(0.12 - x)$$

Newton's Second Law can now be applied giving

$$T_2 - T_1 = m\ddot{x}$$

$$\Rightarrow \quad 50(0.12 - x) - 75(0.08 + x) = 0.5\ddot{x}$$

$$\Rightarrow -125x = 0.5\ddot{x}$$

$$\Rightarrow \quad \ddot{x} = -250x$$

This is the equation of motion for SHM with $\omega = \sqrt{250}$.

The period is $\frac{2\pi}{\omega}$ or 0.40 seconds (to 2 decimal places).

Exercise 3C

In this exercise take g as $9.8\ \text{ms}^{-2}$ (unless otherwise stated).

1. A simple pendulum of length 3 m has a bob of mass 0.2 kg. It is hanging vertically when it is set in motion by a single, sharp sideways blow to the bob which causes the pendulum to oscillate with an angular amplitude of 5°.
   (i) State the amplitude in radians.
   (ii) Calculate the approximate period of the pendulum.
   (iii) The motion of the pendulum can be modelled by

   $$\theta = a\sin(\omega t + \varepsilon)$$

   Write down the values of a, ω and ε.
   (iv) Calculate the maximum speed of the pendulum bob according to this model.

Exercise 3C continued

2. A pendulum has a bob of mass $0.25\,\text{kg}$. The period of the pendulum is 2 seconds. The amplitude of its swing is $3°$.
(i) Find the length of the pendulum.
(ii) State the effect (if any) on the period of the pendulum of:
  (a) making the mass of the bob $0.75\,\text{kg}$;
  (b) doubling the length of the pendulum;
  (c) halving the amplitude of its swing;
  (d) moving it to the moon where the acceleration due to gravity is $1.6\,\text{ms}^{-2}$

3. A pendulum consists of a stone of mass $0.2\,\text{kg}$ attached to a light inelastic string of length $4\,\text{m}$.
(i) Formulate the equation of motion of the stone, stating any assumptions involved.
(ii) Calculate the period of small oscillations.

4. A loaded test tube of total mass m floats in water and is in equilibrium when a length l is submerged as shown. The upward force exerted by the water on the tube is F.

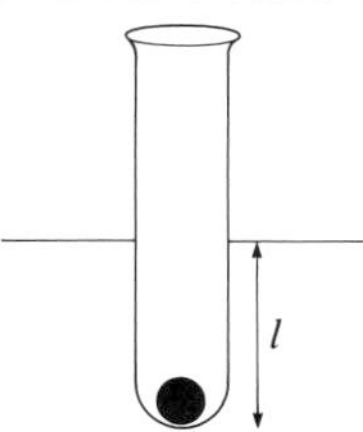

(i) Why is F equal to mg in the equilibrium position? Given that F is directly proportional to the submerged length, find the constant of proportionality in terms of l, m and g.

The test tube is now pushed down a small amount and released.
(ii) Find F when the bottom of the tube is a distance x below its equilibrium position, and use Newton's Second Law to write down the equation of motion for the test tube. Show that it will oscillate with simple harmonic motion with period:

$$T = 2\pi\sqrt{\frac{l}{g}}$$

5. A small block of mass $2\,\text{kg}$ is attached to a fixed point by a spring of stiffness $10\,\text{Nm}^{-1}$ and natural length $0.5\,\text{m}$. The block moves in a straight line on a smooth horizontal table. The block is held in a position such that the spring is stretched $0.2\,\text{m}$, and is then released. At time t seconds later the extension of the spring is x metres.
(i) Show that the equation of motion is given by: $\ddot{x} + 5x = 0$.
(ii) State the initial conditions for the motion in terms of x and t.
(iii) Show that the expression $x = 0.2\cos\sqrt{5}t$ satisfies both parts (i) and (ii).
(iv) State the frequency and amplitude of the block's oscillations.
(v) Find the greatest speed of the block.

Exercise 3C continued

6. Two identical springs of natural length $20\,\text{cm}$ and stiffness $60\,\text{N}\,\text{m}^{-1}$ are attached to a block of mass $200\,\text{g}$ and length $5\,\text{cm}$ which lies on a smooth table. The other ends of the springs are attached to fixed points $45\,\text{cm}$ apart and the springs and the block lie along a straight line as shown in the diagram.

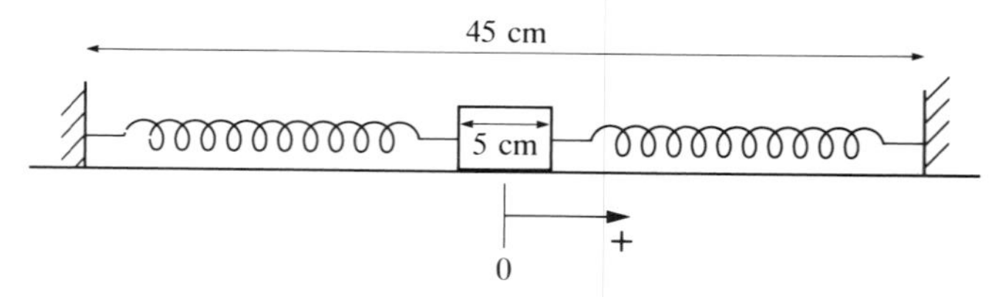

(i) Show that in the equilibrium position there is no tension in either spring.

The block is displaced $2\,\text{cm}$ in the negative direction and released. At time t seconds later, its displacement is x metres.

(ii) Write down the equation of motion of the block.
(iii) Write, in terms of x and t, the initial conditions of the motion.
(iv) Write down an expression for the displacement x at any time during the subsequent motion.
(v) State the frequency of the block's oscillations.
(vi) Find the greatest speed and greatest acceleration of the block.

7. A particle, P, of mass m is placed on a smooth, horizontal track and is attached by two springs to fixed points on the track. The spring AP has natural length $3l$ and modulus of elasticity λ. The spring BP has natural length $2l$ and modulus of elasticity 2λ. The points A and B are a distance $6l$ apart.

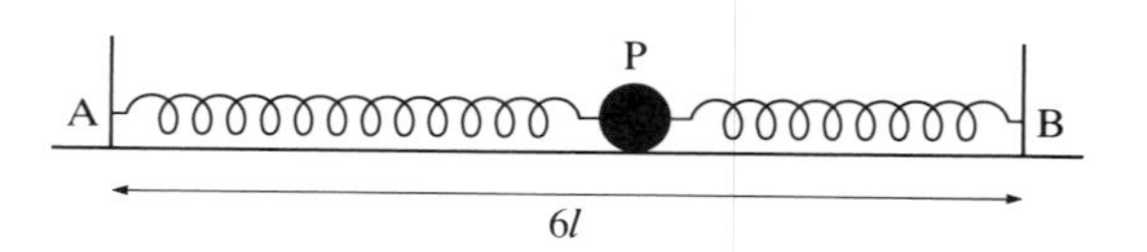

(i) Find the distance of the equilibrium position from A.
(ii) Find the tensions in the two springs when the particle is in equilibrium.

When the particle is stationary at the equilibrium position, it is given an impulse μ in the direction AB. Its subsequent displacement from the equilibrium position is given by x at time t.

(iii) Write down the equation of motion of the particle and its initial conditions.
(iv) Find an expression for x in terms of t.
(v) Write down expressions for the amplitude and frequency of the subsequent oscillations.

8. A block of mass $750\,\text{g}$ hangs vertically, attached to a light elastic string of natural length $50\,\text{cm}$ and modulus of elasticity $150\,\text{N}$. The other end of the string is attached to a ceiling.

(i) Taking g to be $10\,\text{ms}^{-2}$, find the extension of the string in the equilibrium position.

The block is then pulled down a further distance.

(ii) What is the tension in the string when the block is $x\,\text{cm}$ below the equilibrium position?

(iii) Show that when the block is allowed to move, its motion is governed by the equation $\ddot{x} = -400x$.

The block is released from a position $2\,\text{cm}$ below the equilibrium position.

(iv) Find an expression for the displacement $x\,\text{cm}$, t seconds after the block's release.

(v) On another occasion the block is pulled down $3\,\text{cm}$ below the equilibrium position. Explain why in this case the subsequent motion is not simple harmonic.

9. A block of mass m is to be supported by two elastic strings of natural length l_0 and modulus of elasticity λ.

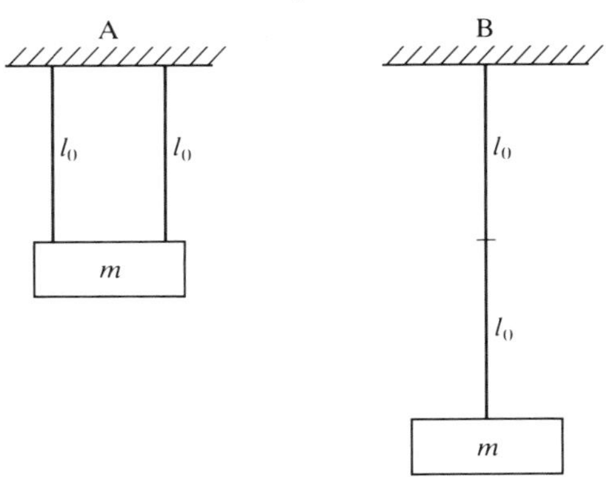

For each of the configurations A and B, find:

(i) the equilibrium position;

(ii) the equation of motion when the block is displaced from the equilibrium position;

(iii) the period of oscillations, assumed small enough for the motion to be simple harmonic, when the block is displaced.

10. A particle of mass $0.02\,\text{kg}$ is performing simple harmonic motion with centre O, period $6\,\text{s}$ and amplitude $2\,\text{m}$.

(i) Find, correct to 2 significant figures in each case:

(a) the maximum speed V of the particle;

(b) the distance of the particle from O when its speed is $0.9\,V$;

(c) the magnitude of the force acting on the particle when it is $1\,\text{m}$ from O.

(ii) Show also that the power exerted by the force on the particle is a periodic function of time and state the period.

Investigation

Grandfather

Here are some lines from a well-known song:

"My grandfather's clock was too tall for the shelf
It was taller by half than the old man himself
Tick, tock, tick, tock, his life's seconds numbering."

Estimate how tall grandfather was.

KEY POINTS

- Motion for which $\ddot{x} = -\omega^2 x$ is simple harmonic.
- The graph illustrates the amplitude and period of simple harmonic motion.

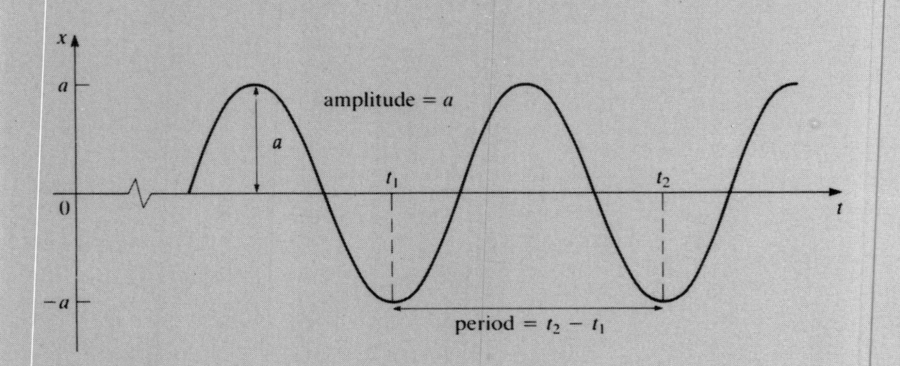

- The speed at displacement x is given by
$$v^2 = \omega^2(a^2 - x^2)$$
- The period is $T = \dfrac{2\pi}{\omega}$
- The frequency is $f = \dfrac{1}{T} = \dfrac{\omega}{2\pi}$

continued

Key Points continued

- **Different ways of expressing SHM:**

Initial conditions	**Displacement**	**Velocity**	**Acceleration**
	x	$v=\dot{x}$	$\ddot{x}$
$x=0$ and $\dot{x}>0$	$a\sin \omega t$	$a\omega\cos \omega t$	$-a\omega^2\sin \omega t$
$x=a$ and $\dot{x}=0$	$a\cos \omega t$	$-a\omega\sin \omega t$	$-a\omega^2\cos \omega t$
Otherwise			
(i)	$a\cos(\omega t+\varepsilon)$	$-a\omega\sin(\omega t+\varepsilon)$	$-a\omega^2\cos(\omega t+\varepsilon)$
(ii)	$a\sin(\omega t+\varepsilon)$	$a\omega\cos(\omega t+\varepsilon)$	$-a\omega^2\sin(\omega t+\varepsilon)$
or (iii)	$A\cos \omega t+B\sin \omega t$ where $a=\sqrt{(A^2+B^2)}$	$-A\omega\sin \omega t+B\omega\cos \omega t$	$-\omega^2(A\cos \omega t+B\sin \omega t)$

For all these functions $x=0$ in the position of equilibrium. If x takes another value, x_0, in the equilibrium position, then $\ddot{x}=-\omega^2(x-x_0)$ and x_0 must be added to functions in the first column.

4 Volumes of revolution and centres of mass by integration

Here's a fine revolution, an' we had the trick to see't

William Shakespeare

What do the shapes of the objects have in common? For each one, suggest a way of finding its volume.
How would you find the volume of
(i) an egg?
(ii) a table tennis ball?
(iii) the moon?

Calculating volumes

One way of finding the volume of an irregular shape is to immerse it in water and measure the apparent increase in the volume of the water. This method is satisfactory for finding the volume of something small and dense like an egg, but you would find it difficult to use for a table tennis ball, and it would be impossible to find the volume of the moon in this way!

You will be familiar with the formula $V = \frac{4}{3}\pi r^3$ for the volume of a sphere. The methods you meet in this chapter will enable you to prove this formula, and also to calculate the volumes of other solids with *rotational symmetry* like the ones illustrated above.

Volumes of revolution

If a set-square (without a hole) is rotated through 360° about one of its shorter sides its face will sweep out a solid cone (figure 4.1(a)). A 60° set-square will sweep out different cones depending on the side chosen. In the same way a semi-circular protractor will sweep out a sphere if it is rotated completely about its diameter as shown in figure 4.1(b).

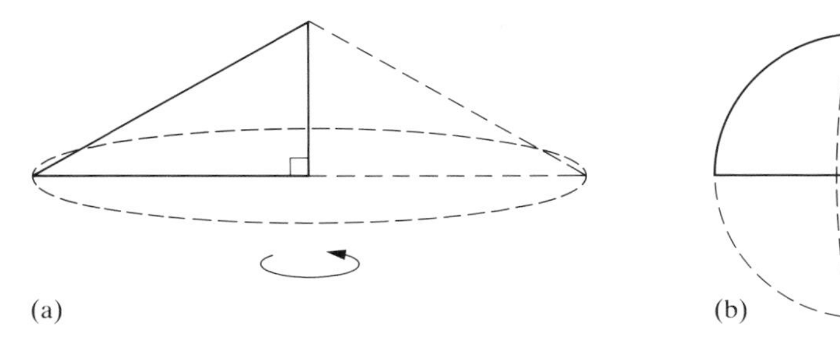

Figure 4.1

The cone and sphere are examples of solids of revolution, as are the objects illustrated at the beginning of this chapter. If any region of a plane is rotated completely about an axis in the plane it will form a solid of revolution. Any solid body with rotational symmetry can be formed in this way.

Imagine that such a solid is cut into thin slices perpendicular to the axis of rotation just as a potato is sliced to make crisps (figure 4.2). Each slice will approximate to a thin disc, but the discs will vary in radius. The volume of the solid can be found by adding the volumes of these discs and then finding the limit of the volume as their thickness approaches zero. When thin discs are used to find a volume they are called *elementary discs*. (Similarly when thin strips are used to find an area they are called *elementary strips*.)

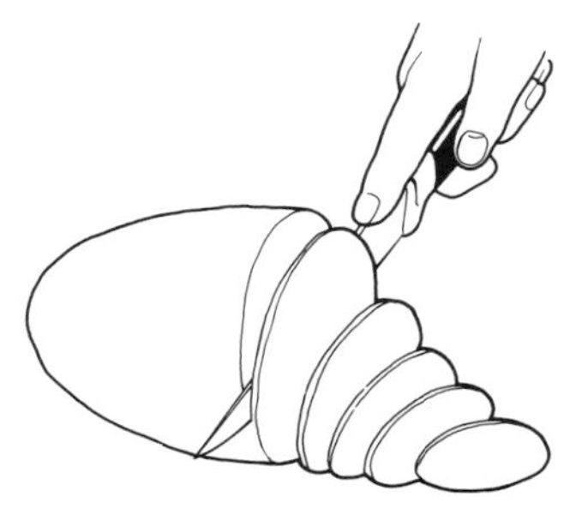

Figure 4.2

This is essentially the same procedure as that developed for finding the area under a curve (*Pure Mathematics 1*, chapter 6), and so, as then, integration is required.

Volumes of revolution about the x axis

Look at the solid formed by the rotation about the x axis of the region under the graph of $y = f(x)$ between the values $x = a$ and $x = b$ as shown in figure 4.3. (You should always start by drawing diagrams when finding areas and volumes.)

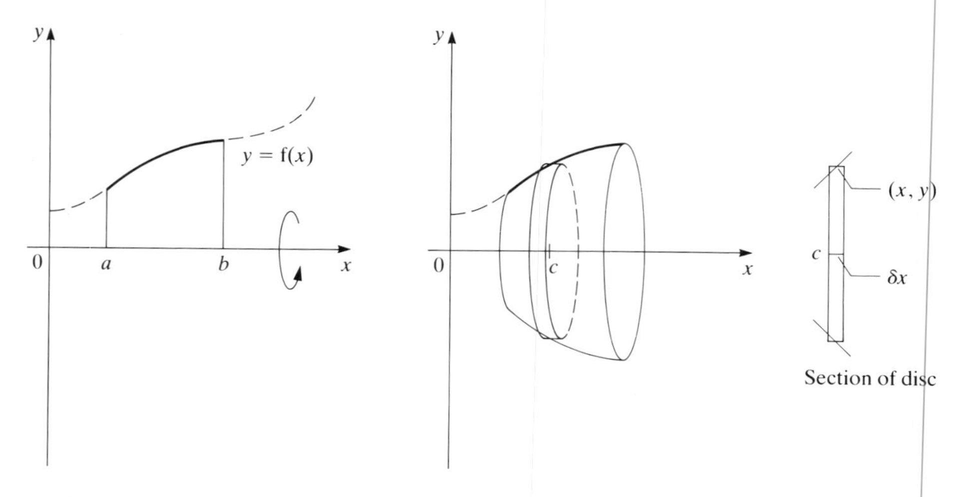

Figure 4.3

The volume of the solid of revolution (which is usually called the volume of revolution) can be found by imagining that it can be sliced into thin discs.

Each disc is a thin cylinder of radius y and thickness δx, so its volume is given by

$$\delta V = \pi y^2 \delta x$$

The volume of the solid is the limit of the sum of all these elementary volumes as $\delta x \to 0$ i.e. the limit as $\delta x \to 0$ of

$$\sum_{\substack{\text{over all}\\ \text{discs}}} \delta V = \sum_{x=a}^{x=b} \pi y^2 \delta x$$

The limiting values of sums such as these are integrals so

$$V = \int_a^b \pi y^2 \mathrm{d}x$$

The limits are a and b because x takes values from a to b.

It is essential to realise that this integral is written in terms of dx and so it cannot be evaluated unless the function y is also written in terms of x. For this reason the integral is often written as

$$V = \int_a^b \pi[\mathrm{f}(x)]^2 \, \mathrm{d}x.$$

EXAMPLE

Find the volume of revolution formed when the region between the curve $y = x^2 + 2$, the x axis and the lines $x = 1$ and $x = 3$ is rotated completely about the x axis.

Solution

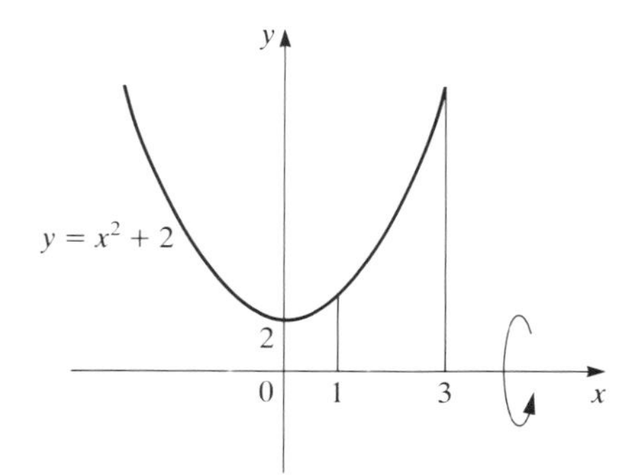

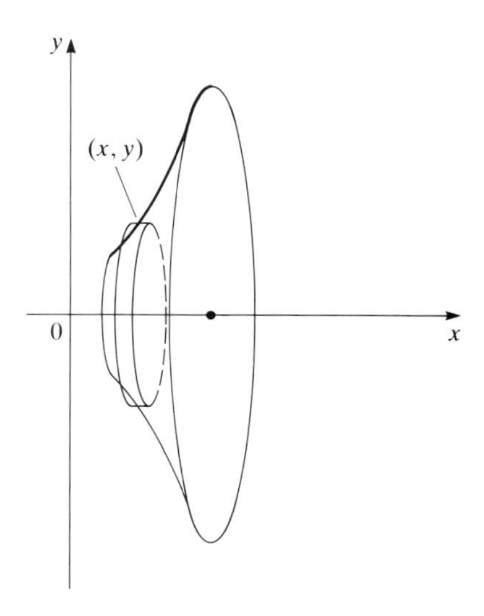

The volume is:

$$V = \int_a^b \pi[f(x)]^2 \, dx$$

$$= \int_1^3 \pi(x^2 + 2)^2 \, dx$$

$$= \int_1^3 \pi(x^4 + 4x^2 + 4) \, dx$$

$$= \left[\pi\left(\frac{x^5}{5} + \frac{4x^3}{3} + 4x\right)\right]_1^3$$

$$= \pi\left[\left(\frac{3^5}{5} + 36 + 12\right) - \left(\frac{1}{5} + \frac{4}{3} + 4\right)\right]$$

The volume is $\frac{1366\pi}{15}$ cubic units or 286 cubic units correct to 3 significant figures.

NOTE

Unless a decimal answer is required, it is usual to leave π in the answer, which is then exact.

EXAMPLE Find the volume of a spherical ball of radius 2 cm.

Solution

The sphere is formed by rotating the circle $x^2 + y^2 = 4$ about the x axis.

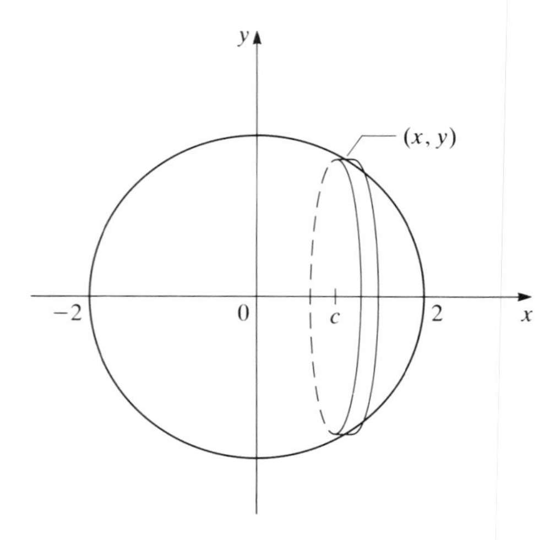

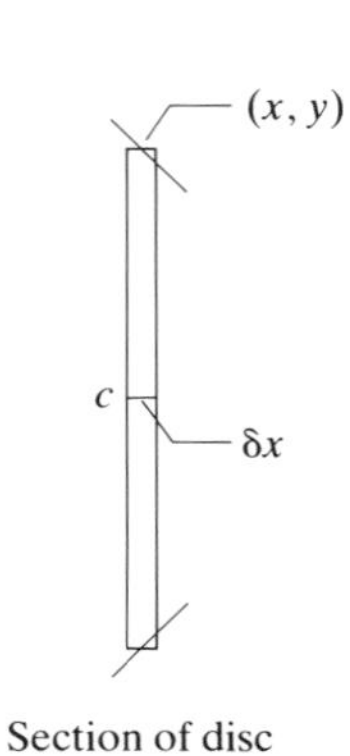

Section of disc

The volume of the sphere is found by using

$$V = \int_a^b \pi y^2 \, dx$$

with limits $a = -2$ and $b = 2$.

Since $x^2 + y^2 = 4$, then $y^2 = 4 - x^2$, and the integral may be written as

$$V = \int_{-2}^{2} \pi(4 - x^2) \, dx.$$

The constant π can be taken outside the integral so we have

$$V = \pi\left[4x - \frac{x^3}{3}\right]_{-2}^{2}$$

$$= \pi\left[\left(8 - \frac{8}{3}\right) - \left(-8 + \frac{8}{3}\right)\right]$$

$$= \frac{32\pi}{3} = 33.51\ldots$$

So the volume of a spherical ball with radius 2 cm is 33.5 cm^3 correct to 3 significant figures.

The volume of a sphere

The formula for the volume of a sphere of radius r can be found by following the same argument, but using a circle of radius r rather than one of radius 2 cm. Its equation is $x^2 + y^2 = r^2$ and the limits are $-r$ and r.

$$V = \int \pi y^2 \mathrm{d}x$$

$$= \int_{-r}^{r} \pi(r^2 - x^2)\,\mathrm{d}x$$

Therefore

$$V = \pi\left[r^2x - \frac{x^3}{3}\right]_{-r}^{r}$$

$$= \pi\left[\left(r^3 - \frac{r^3}{3}\right) - \left(-r^3 - \frac{-r^3}{3}\right)\right]$$

$$\Rightarrow \quad V = \frac{4}{3}\pi r^3$$

Volumes of revolution about the y axis

If the same portion of the curve $y = x^2 + 2$ were to be used to form a region which could be rotated about the y axis, a very different solid would be obtained from that on p. 115. The region required would be that which lies between the curve and the y axis and the limits $y = 3$ and $y = 11$ shown in figure 4.4.

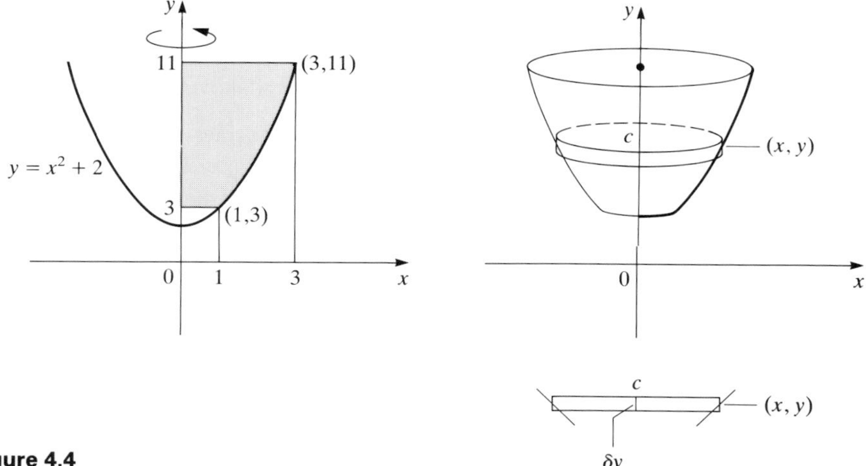

Figure 4.4

An elementary disc has an approximate volume $\delta V = \pi x^2 \delta y$ and the volume of revolution about the y axis is given by:

$$V = \int \pi x^2 \mathrm{d}y$$

This time it is necessary to write x^2 in terms of y. In this case $x^2 = y - 2$ and so the final integral becomes:

$$V = \int_3^{11} \pi(y-2)\,dy$$

$$V = \left[\pi\left(\frac{y^2}{2} - 2y\right)\right]_3^{11}$$

$$\Rightarrow \quad V = \pi\left[\left(\frac{11^2}{2} - 22\right) - \left(\frac{3^2}{2} - 6\right)\right]$$

The volume is 40π cubic units.

Finding a mathematical model for a solid of revolution

Experiment

Use an immersion method to find the volumes of two eggs of different shapes or sizes. Measure their lengths and widths as shown in the diagram.

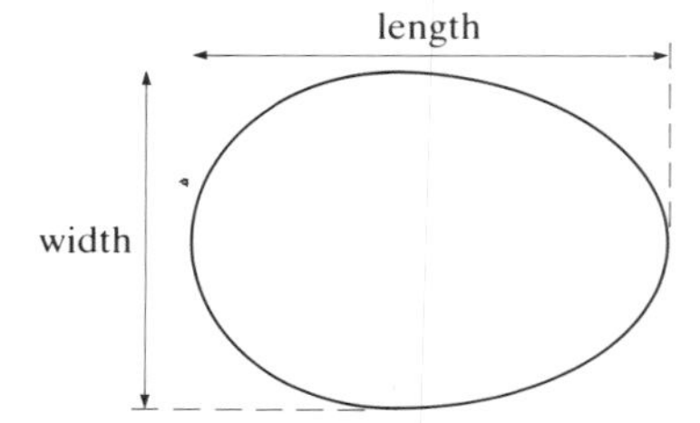

We are now going to construct a mathematical model to find the volume of an egg. You will then be able to use your measurements to check the accuracy of this model.

A hen's egg has rotational symmetry about its longest axis so it can be represented by a solid of revolution. But what curve should be revolved around the x axis to obtain a reasonable egg shape?

The cross-section looks circular at the "blunt" end, but the "sharp" end is longer and looks more like half an ellipse. It may be possible to model the egg as part of a circle plus part of an ellipse, rotated about the x axis. The equation of an ellipse of length $2a$ and width $2b$ as in figure 4.5 can be written as:

$$\frac{x^2}{a^2} + \frac{y^2}{b^2} = 1$$

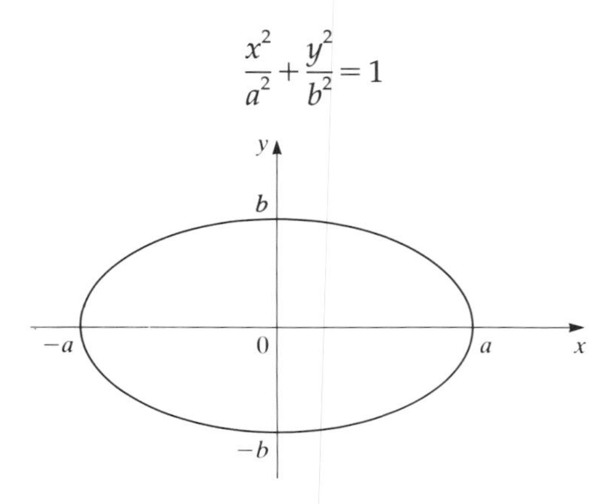

Figure 4.5

A particular egg is 6.3 cm long and 4.6 cm wide. Figure 4.6 shows the regions which are rotated about the x axis.

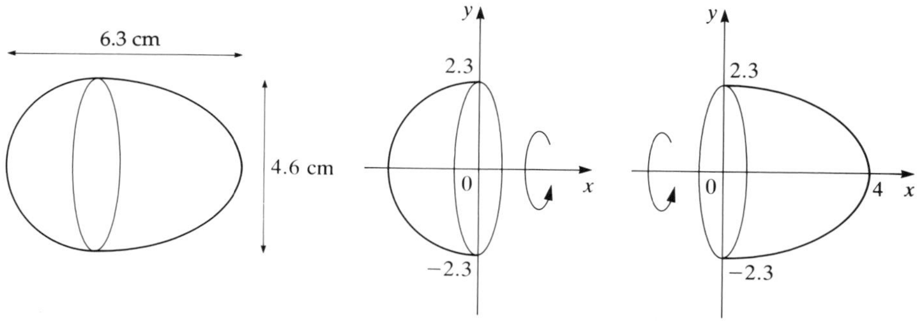

Figure 4.6

The left part is a hemisphere of radius 2.3 cm so its volume is:

$$V_1 = \tfrac{2}{3}\pi r^3 = \tfrac{2}{3}\pi(2.3)^3$$
$$= 25.48\,\text{cm}^3 \text{ (to 2 decimal places)}$$

The equation of the elliptical part to the right is

$$\frac{x^2}{4^2} + \frac{y^2}{2.3^2} = 1$$

On rearrangement this becomes

$$y^2 = 2.3^2\left(1 - \frac{x^2}{4^2}\right)$$

The volume of the elliptical part is therefore given by

$$V_2 = \int_0^4 \pi y^2\,\mathrm{d}x$$
$$= \pi \times 2.3^2 \int_0^4 \left(1 - \frac{x^2}{4^2}\right)_0^4$$
$$= \pi \times 2.3^2 \left[x - \frac{x^3}{48}\right]_0^4$$
$$= \pi \times 2.3^2 \left[\left(4 - \frac{4^3}{48}\right) - 0\right]$$
$$= 44.32\,\text{cm}^3 \text{ (to 2 decimal places).}$$

The total volume of the egg can now be found as

$$V_1 + V_2 = 25.48 + 44.32$$
$$= 70\,\text{cm}^3 \text{ (correct to 2 significant figures)}$$

Activity

Use this method to calculate the volume of one of the eggs you measured in the experiment, then use the actual volume (found by immersing the egg in water) to check the model against reality.

Exercise 4A

1. In each part of this question a region is defined in terms of the lines which form its boundaries. Draw a sketch of the region and find the volume of the solid obtained by rotating it through $360°$ about the x axis.

(a) $y=2x$, the x axis and the lines $x=1$ and $x=3$.

(b) $y=\sqrt{x}$, the x axis and the line $x=4$.

(c) $y=2x+1$, the x axis and the lines $x=1$ and $x=5$.

(d) $y=\dfrac{1}{x}$, the x axis and the lines $x=1$ and $x=3$.

(e) $y=3e^x$, the x axis, the y axis and the line $x=1.5$.

(f) $y=\cos x$ (for $0 \leqslant x \leqslant \frac{\pi}{2}$), the x axis and the y axis. (Hint: $\cos^2 x=\frac{1}{2}(1+\cos 2x)$).

2. In each part of this question a region is defined in terms of the lines which form its boundaries. Draw a sketch of the region and find the volume of the solid obtained by rotating it through $360°$ about the y axis.

(a) $y=x^3$, the y axis and the lines $y=0$ and $y=8$.

(b) $y=2x-1$, the y axis and the line $y=5$.

(c) $x^2+y^2=9$, the y axis and the lines $y=1$ and $y=3$.

(d) $\dfrac{x^2}{9}+\dfrac{y^2}{4}=1$ and the y axis.

3. (i) Sketch the graph of $x^2+y^2=36$.

(ii) A hemispherical bowl of internal radius $6\,\text{cm}$ is filled with water to a depth of $4\,\text{cm}$. Find the volume of water in litres (1 litre $=1000\,\text{cm}^3$).

4. A mathematical model for a large garden pot is obtained by rotating through $360°$ about the y axis the part of the curve $y=0.0001x^4$ which is between $x=10$ and $x=25$, and then adding a flat base. Units are in centimetres.

(i) Draw a sketch of the curve and shade in the cross-section of the pot, indicating which line will form its base.

(ii) Garden compost is sold in litres. How many litres will be required to fill the pot to a depth of $35\,\text{cm}$? (Ignore the thickness of the pot.)

5. The triangle OAB shown in the diagram is rotated about the x axis to form a solid cone.

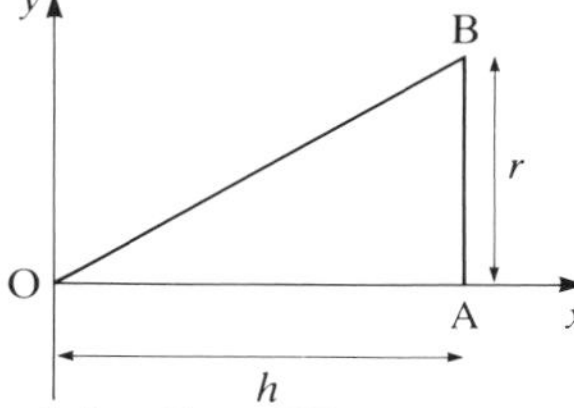

   (i) Find the equation of the line OB.
   (ii) Hence show that the volume of a cone with radius r and height h is $\frac{1}{3}\pi r^2 h$.

6. (i) Sketch the ellipse $\frac{x^2}{a^2}+\frac{y^2}{b^2}=1$ where $a>b$.

   The region under this curve and above the x axis is rotated through 360° about the x axis to form an ellipsoid.
   (ii) Draw a diagram to illustrate the solid formed.
   (iii) Show that the volume of this solid is $\frac{4}{3}\pi b^2 a$.
   A mathematical model for an egg is that one end is a half ellipsoid of length a and greatest radius b and the other end is a hemisphere of radius b.
   (iv) Write down the volume of the egg in terms of a and b.

7. (i) Draw a diagram to illustrate the solid formed when the ellipse $\frac{x^2}{a^2}+\frac{y^2}{b^2}=1$ $(a>b)$ is rotated about the y axis.

The photograph shows the elliptical galaxy M87 seen edge on. This galaxy is estimated to have a maximum diameter of 10 000 light years.

   (ii) Estimate its volume in metres cubed. (You will need to take measurements from the photograph. The speed of light is approximately $3\times10^8\,\mathrm{ms}^{-1}$.)

8. A cod liver oil capsule can be modelled by an ellipsoid which is produced by rotating an ellipse of length 1.5 cm and width 0.9 cm about its major axis. The capsule contains 0.32 ml of cod liver oil. What proportion of the volume of the capsule is this?

Exercise 4A continued

9. The diagram shows the hyperbola $\frac{x^2}{a^2}-\frac{y^2}{b^2}=1$.

When it is rotated through $360°$ about the y axis a hyperboloid is formed.

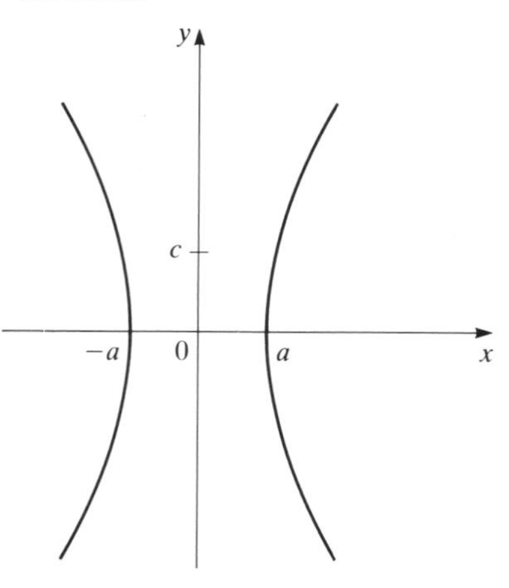

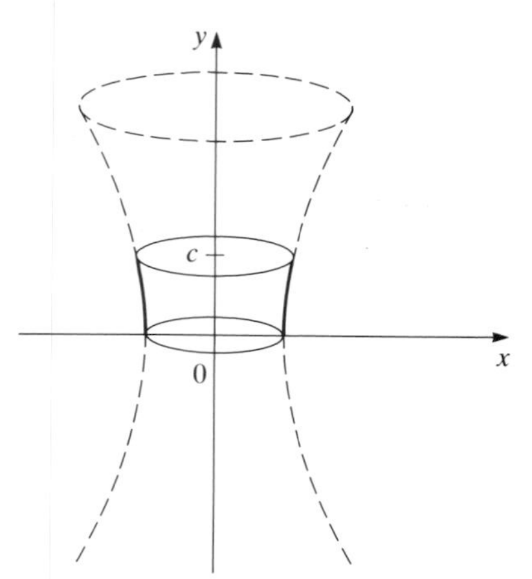

(i) Show that the volume enclosed by this hyperboloid between $y=0$ and $y=c$ is given by: $V=\frac{\pi a^2 c}{3b^2}(3b^2+c^2)$.

The cooling towers of power stations can be modelled as hyperboloids and the dimensions of one such tower, at Didcot power station are shown on the picture.

A = 51·3 m

B = 47·5 m

C = 24·5 m

D = 89·5 m

E = 85·5 m

(ii) Verify that the equation $\frac{x^2}{560}-\frac{y^2}{3600}=1$ is a suitable mathematical model for the curve which is rotated to form the hyperboloid.

(iii) Use the formula above to find the volume occupied by the cooling tower, and show that the six towers of the power station occupy over 1.9 million cubic metres of space.

10. A mathematical model for the curved surface of a measuring jug is obtained by rotating the part of the curve $y = 0.003x^4$ between $y = 1$ and $y = 10$ through $360°$ about the y axis. It has a flat base. The units are centimetres.

(i) Sketch the jug.

(ii) Find the values of y for which marks must be placed on the side of the jug to indicate volumes of

a) 100 ml b) 0.5 litre c) 1 litre

11. The circle $x^2 + (y+1)^2 = 5$ meets the x axis at the points A and B. The minor arc AB is rotated about the x axis to form a solid of revolution which can be used to model a rugby ball.

(i) Find the coordinates of A and B and show that

$$y^2 = 6 - x^2 - 2\sqrt{(5 - x^2)}$$

$$\left(\text{Hint: } \int_0^2 \sqrt{(5 - x^2)}\, dx = \frac{5}{2}\arcsin\frac{2}{\sqrt{5}} + 1\right)$$

(ii) Hence calculate the volume of the solid of revolution.

(iii) A similar rugby ball is 30 cm long. Find its volume.

Investigations

Ceramics

It is possible to use the trapezium rule to find an approximation to any definite integral which cannot be solved analytically or for which an algebraic formula is not known.

Above is a photograph of a group of ceramics made by the potter Hans Coper. Given that the height of the front form is 0.5 m, how could you find its external volume using measurements taken from the photograph?

Investigation continued

Hot air balloon

It has been said that a hot air balloon contains about a ton of air. Can this be true?

The moon

How can the volume of the moon be estimated using only measurements which can be made by a student and Newton's law of gravitation? (The law states that the force of attraction between bodies of mass m_1 and m_2 whose centres of mass are a distance r apart is given by $\frac{Gm_1m_2}{r^2}$ where G is a constant.)

You might need to consider the period of the moon's rotation about the earth, the value of g at the earth's surface and the radius of the earth in order to decide how far the moon is from you. Could you find all these? A meter rule and a small button should then enable you to find the diameter of the moon, and therefore its volume.

Centres of mass

The position of the centre of mass of a solid of revolution will affect its stability and the way it will roll. Some toys are designed so that they can never be knocked over. Many birds that live on cliffs lay pointed eggs which roll round in circles so that they do not fall over the cliff edge. The position of the centre of mass is also important in the design of a lot of sporting equipment.

Using integration to find centres of mass

The calculus methods you have been using to determine the volumes of solids of revolution can be extended to find their centres of mass (assuming they are of uniform density).

Notice that because of symmetry, the centre of mass of a solid of revolution must lie on its axis, so provided you choose either the x axis or the y axis, to be the axis of symmetry there is only one co-ordinate to determine. Take a solid of revolution about the x axis, and divide the solid into thin discs as before. For an elementary disc situated at the point (x,y) on the curve, the centre of mass is at the point $(x,0)$ on the x axis as shown in figure 4.7.

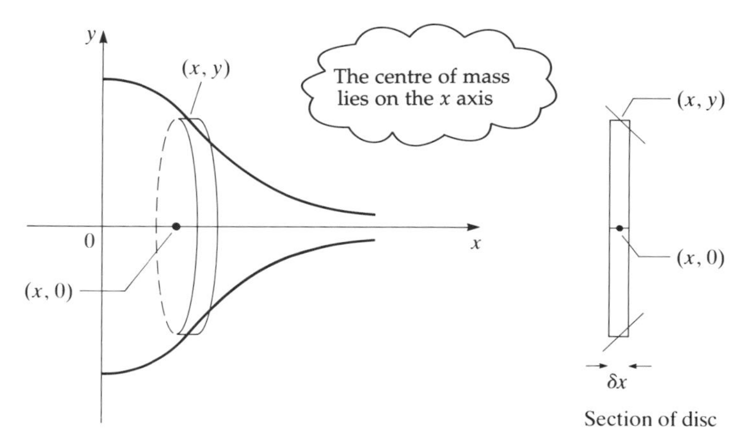

Figure 4.7

The volume of this disc is $\delta V = \pi y^2 \, \delta x$

and so its mass is $\delta M = \rho \pi y^2 \, \delta x$

where ρ is the density.

The solid is now approximated by the sum of such discs. For each of these both its mass and the position of its centre of mass are known, so you can use the result from *Mechanics 2*, chapter 3. For a composite body, the position, $\bar{x}$, of the centre of mass is given by

Moment of whole mass at centre of mass = Sum of moments of individual masses

$$M\bar{x} = \sum_i m_i x_i$$

In this case:

$$\left(\sum_{\substack{\text{All} \\ \text{discs}}} \delta M \right) \bar{x} = \sum_{\substack{\text{All} \\ \text{discs}}} (\delta M x)$$

Substituting the expression for δM obtained above gives:

$$(\Sigma \rho \pi y^2 \, \delta x)\bar{x} = \Sigma \rho \pi y^2 x \delta x$$

In the limit as $\delta x \to 0$ these sums may be represented by integrals and so

$$(\int \rho \pi y^2\,dx)\bar{x} = \int \rho \pi y^2 x\,dx$$

Assuming that ρ is uniform, you can divide through by $\rho\pi$ to give

$$(\int y^2\,dx)\bar{x} = \int y^2 x\,dx \quad \text{or} \quad \bar{x} = \frac{\int y^2 x\,dx}{\int y^2\,dx}$$

This result is valid for any solid of revolution of *uniform density* which is formed by rotation about the x axis.

This result may also be written as

$$V\bar{x} = \int \pi y^2 x\,dx$$

where V is the volume of the solid.

In the next example, the centre of mass of a solid hemisphere is found using the above result.

EXAMPLE

Find the centre of mass of a solid hemisphere of radius r.

Solution

With the hemisphere oriented as in the diagram, the point (x,y) lies on the curve $x^2 + y^2 = r^2$ and the limits for the integration are $x = 0$ and $x = r$. We also know that the volume is $\frac{2}{3}\pi r^3$ but the equations will look simpler if this is substituted later.

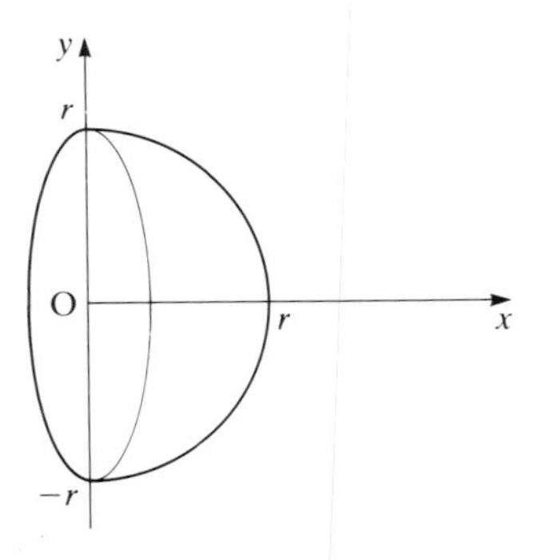

$$V\bar{x} = \int_0^r \pi y^2 x\,dx$$

Substituting $y^2 = r^2 - x^2$ gives

$$V\bar{x} = \int_0^r \pi(r^2 - x^2)x\,dx$$

$$= \int_0^r \pi(r^2 x - x^3)\,dx$$

$$= \pi\left[\frac{r^2x^2}{2} - \frac{x^4}{4}\right]_0^r$$

So $V\bar{x} = \frac{1}{4}\pi r^4$

Substituting for V gives: $\frac{2}{3}\pi r^3\bar{x} = \frac{1}{4}\pi r^4$

$\Rightarrow \quad \bar{x} = \frac{3}{8}r$

The centre of mass of a solid hemisphere is $\frac{3}{8}r$ from the centre of its base.

When finding centres of mass in this way it is often necessary to use *integration by parts* (see *Pure Mathematics 3*, chapter 3). The following example illustrates this.

EXAMPLE

Find the position of the centre of mass of the solid of revolution formed when the region between the curve $y = 2e^x$, the lines $x = 0$ and $x = 2$, and the x axis is rotated through 360° about the x axis.

Solution

The diagrams show the region and the solid obtained.

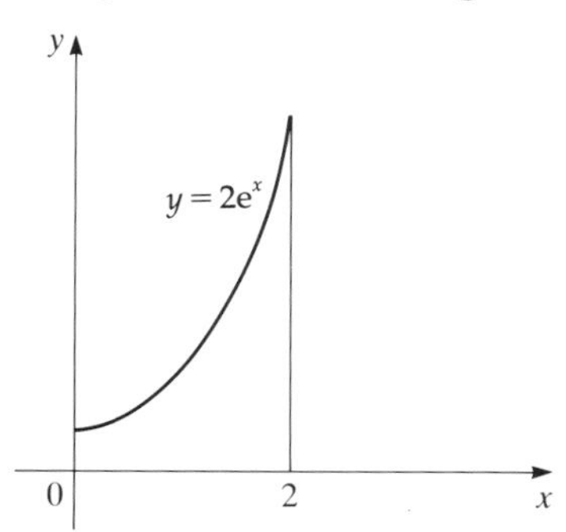

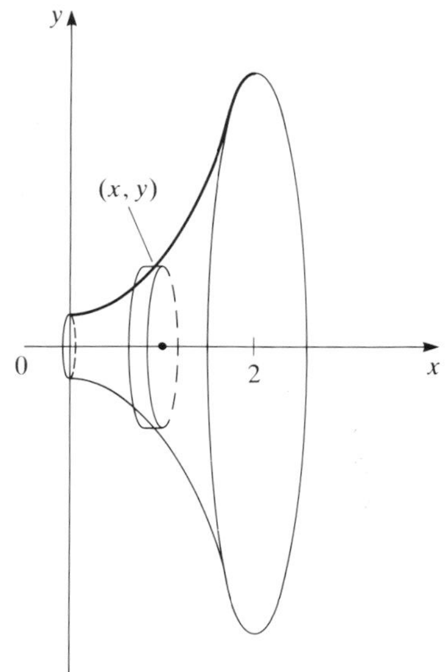

The volume, V, of the solid is given by

$$V = \int \pi y^2 \, dx$$

$$= \pi \int_0^2 (2e^x)^2 \, dx$$

$$= 4\pi \int_0^2 e^{2x} \, dx$$

$$= 4\pi \left[\frac{e^{2x}}{2} \right]_0^2$$

$$= 2\pi(e^4 - 1).$$

By symmetry, the y co-ordinate of the centre of mass is 0. The x co-ordinate is given by $\bar{x}$ in

$$V\bar{x} = \int \pi y^2 x \, dx$$
$$= \pi \int_0^2 (2e^x)^2 x \, dx$$
$$= 4\pi \int_0^2 xe^{2x} \, dx. \qquad ①$$

This integral requires the technique of integration by parts (see *Pure Mathematics 3*, chapter 3), expressed by the general formula

$$\int_a^b u\frac{dv}{dx}\,dx = [uv]_a^b - \int_a^b v\frac{du}{dx}\,dx.$$

To find $\int xe^{2x}\,dx$,

let $u = x \Rightarrow \frac{du}{dx} = 1$,

and let $\frac{dv}{dx} = e^{2x} \Rightarrow v = \frac{e^{2x}}{2}$.

The limits are $a = 0$ and $b = 2$.

Substituting these in the general formula, you obtain

$$\int_0^2 xe^{2x}\,dx = \left[x\frac{e^{2x}}{2}\right]_0^2 - \int_0^2 \frac{e^{2x}}{2} \times 1 \times dx$$
$$= \left[\tfrac{1}{2}xe^{2x}\right]_0^2 - \left[\tfrac{1}{4}e^{2x}\right]_0^2$$
$$= \left(e^4 - 0\right) - \left(\tfrac{1}{4}e^4 - \tfrac{1}{4}\right)$$
$$= \tfrac{1}{4}(3e^4 + 1).$$

Substituting this and $V = 2\pi(e^4 - 1)$ in ①,

$$V\bar{x} = 4\pi \int_0^2 xe^{2x}\,dx$$

gives $$2\pi(e^4 - 1)\bar{x} = 4\pi \times \tfrac{1}{4}(3e^4 + 1)$$

$$\Rightarrow \bar{x} = \frac{(3e^4 + 1)}{2(e^4 - 1)}$$
$$= 1.54 \text{ (correct to 2 decimal places).}$$

The centre of mass is at (1.54, 0).

NOTE *If the answer is required in decimal form, it is best to substitute the numerical value of constants such as e and π later rather than sooner. Notice that π cancels in this example.*

Centres of mass of composite bodies

Once the positions of the centres of mass of standard solids have been found, it is possible to use these results to find the positions of the centres of mass of composite bodies. Some useful results are summarised in the following table.

Diagram	Body	Volume	Height of c.o.m. above base
	Solid sphere, radius r	$\frac{4}{3}\pi r^3$	r
	Solid hemisphere, radius r	$\frac{2}{3}\pi r^3$	$\frac{3}{8}r$
	Solid cone, height h, radius r	$\frac{1}{3}\pi r^2 h$	$\frac{1}{4}h$

You may also find it helpful to know the position of the centre of mass for some shells (hollow bodies with negligible wall thickness), and these are given below. A method for finding these is given in Mathematical Note 3 (page 161).

Diagram	Body	Curved surface area	Height of c.o.m. above base
	Hollow hemisphere, radius r	$2\pi r^2$	$\frac{1}{2}r$
	Hollow cone, height h, radius r	πrl $(l = \sqrt{(r^2 + h^2)})$	$\frac{1}{3}h$

EXAMPLE Find the position of the centre of mass of a uniform hemispherical bowl of thickness 1 cm and inside radius 9 cm.

Solution

Think of the bowl as a solid hemisphere of radius 10 cm from which another solid hemisphere of radius 9 cm has been removed. Then the original 10 cm hemisphere can be treated as a composite body, consisting of the shell and the 9 cm hemisphere.

The 10 cm hemisphere = The bowl + The 9 cm hemisphere

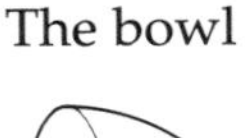

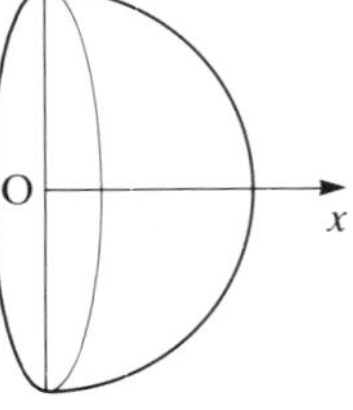

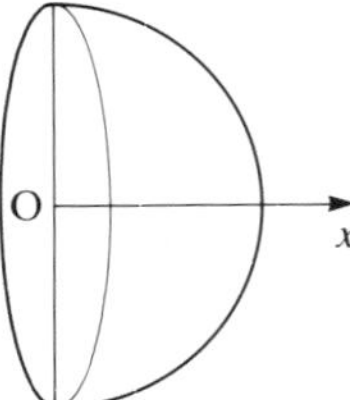

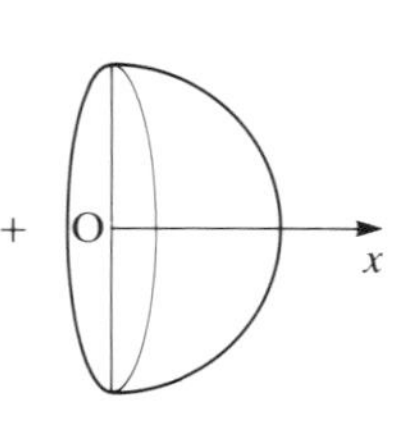

	The 10 cm hemisphere	The bowl	The 9 cm hemisphere
Mass	$\frac{2}{3}\pi\rho 10^3$	$\frac{2}{3}\pi\rho(10^3 - 9^3)$	$\frac{2}{3}\pi\rho 9^3$
Distance from O to c.o.m.	$\frac{3}{8}\times 10$	$\bar{x}$	$\frac{3}{8}\times 9$

Taking moments gives

$$\left(\frac{2}{3}\pi\rho 10^3\right)\times\left(\frac{3}{8}\times 10\right) = \left(\frac{2}{3}\pi\rho(10^3 - 9^3)\times\bar{x}\right) + \left(\frac{2}{3}\pi\rho 9^3\right)\times\left(\frac{3}{8}\times 9\right)$$

This can be simplified by dividing through by $\frac{2}{3}\pi\rho$ to give

$$\frac{3}{8}\times 10^4 = (10^3 - 9^3)\bar{x} + \frac{3}{8}\times 9^4$$

$$\bar{x} = \frac{3(10^4 - 9^4)}{8(10^3 - 9^3)}$$

$$= 4.76 \qquad \text{(to 3 significant figures).}$$

NOTE

An alternative, but equivalent method is to regard the bowl as the sum of a solid hemisphere of radius 10 cm and a solid hemisphere of radius 9 cm which has negative mass.

In this case the moments equation would be written

$$\frac{2}{3}\pi\rho(10^3 - 9^3)\bar{x} = \left(\frac{2}{3}\pi\rho 10^3\right)\times\left(\frac{3}{8}\times 10\right) + \left(-\frac{2}{3}\pi\rho 9^3\right)\times\left(\frac{3}{8}\times 9\right)$$

Exercise 4B

1. In Exercise 4A question 1, you found the volumes of several solids of revolution produced by rotating certain regions through 360° about the x axis. These regions are described again below and the volumes of the resulting solids of revolution are given. In each case:
   (i) Draw a diagram of each solid showing an elementary disc at the point (x,y) on the curve.
   (ii) Determine the position of its centre of mass by equating $V\bar{x}$ to a suitable integral.

   (a) $y = 2x$, the x axis and the lines $x = 1$ and $x = 3$ ($V = \frac{104\pi}{3}$).

   (b) $y = \sqrt{x}$, the x axis and the line $x = 4$ ($V = 8\pi$).

   (c) $y = 2x + 1$, the x axis and the lines $x = 1$ and $x = 5$ ($V = \frac{652\pi}{3}$).

   (d) $y = \frac{1}{x}$, the x axis and the lines $x = 1$ and $x = 3$ ($V = \frac{2\pi}{3}$).

   (e) $y = 3e^x$, the x axis, the y axis and the line $x = 1.5$ ($V = \frac{9\pi}{2}(e^3 - 1)$).

   (f) $y = \cos x$ (from $x = 0$ to $\frac{\pi}{2}$), the x axis and the y axis ($V = \frac{\pi^2}{4}$).

2. In Exercise 4A question 2, you found the volumes of several solids of revolution produced by rotating certain regions through 360° about the y axis. Some of these are described again below and their volumes are given. In each case:
   (i) Draw a diagram of each solid showing an elementary disc at the point (x, y) on the curve.
   (ii) Determine the position of its centre of mass by equating $V\bar{y}$ to a suitable integral.

   (a) $y = x^3$, the y axis and the lines $y = 0$ and $y = 8$ ($V = \frac{96\pi}{5}$)

   (b) $y = 2x - 1$, the y axis and the line $y = 5$ ($V = 18\pi$)

   (c) $x^2 + y^2 = 9$, the y axis and the lines $y = 1$ and $y = 3$ ($V = \frac{28\pi}{3}$)

3. (i) Sketch the curve $y = x^2(2 - x)$.
   (ii) Find the volume of the solid formed when the closed region between the curve and the x axis is rotated about this axis.
   (iii) Find the position of the centre of mass of this solid.

4. (i) Sketch the curve $y^2 = 4x$ (i.e. $y = \pm 2\sqrt{x}$) and shade the region between the curve, the x axis and the lines $x = 1$ and $x = 4$.
   (ii) Find the volume and the position of the centre of mass of the solid generated when the shaded region is rotated completely about the x axis.

Exercise 4B continued

(iii) The region between the same portion of the curve and the y axis is to be rotated through $360°$ about the y axis. Find

(a) appropriate limits for y;

(b) the volume of revolution so formed;

(c) the position of the centre of mass of this new solid of revolution.

5. (i) Sketch the ellipse $\frac{x^2}{9}+\frac{y^2}{4}=1$ and shade the region lying between the curve and the axes for positive values of x and y.

(ii) Find the volume and centre of mass of the solid obtained by rotating the shaded region through $360°$ about the x axis.

6. Water fills a light hemispherical bowl of radius $12\,\text{cm}$ to a depth of $6\,\text{cm}$. Find the height of the centre of mass of the water above the base of the bowl.

7. A toy clown is modelled by a solid cone of radius r and height h, attached to a solid hemisphere of radius r and centre O as shown.

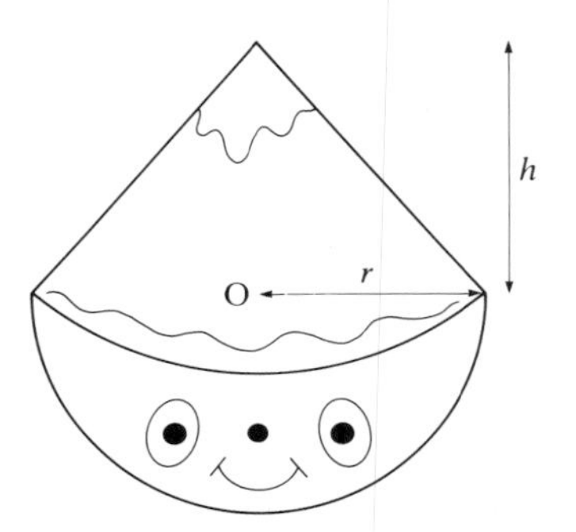

(i) Copy and complete the following table using the results on page 129

	Clown =	Hemisphere +	Cone
Mass			
Distance of c.o.m. above O	$\bar{y}$		

(ii) Write down, but do not simplify, an equation for finding $\bar{y}$.

(iii) If the centre of mass is at the centre of the hemisphere, use your equation to find the ratio of h to r.

(iv) If the ratio of h to r is less than that obtained in part (iii), what position will the toy assume when placed with a point on the hemisphere in contact with the ground?

8. The diagram shows a container consisting of a hollow cylinder of radius 3 cm and height 18 cm with a circular base, centre O, topped by a hollow cone of the same radius and height 4 cm. The container is made of thin, uniform material. The origin O is at the centre of the base.

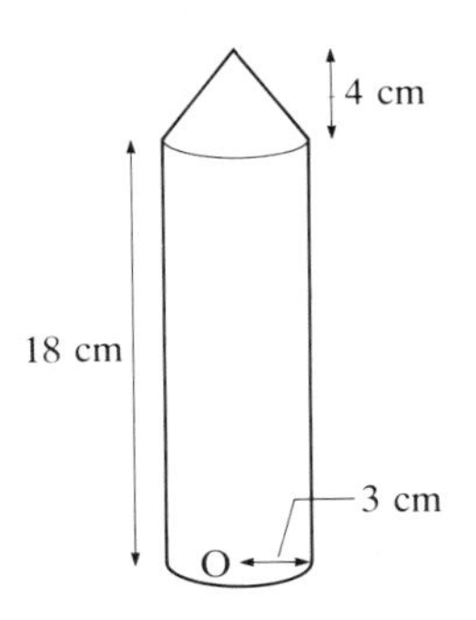

(i) Copy and complete the following table using the results on page 129

	Whole container	=	Cylinder base	+	Cylinder walls	+	Cone
Mass							
Distance of c.o.m. above O	$\bar{y}$						

(ii) Find the height of the centre of mass of the container above the base.

(iii) The container is placed with its circular base on a flat surface in a boat and is prevented from slipping. If the boat heels over so that the surface makes an angle α to the horizontal, find at what value of α the empty container will topple over.

(iv) If the container is filled with liquid to the top of the cylinder, is it more or less likely to topple over than when it is empty?

9. The shape of the bowl of a wine glass is produced by rotating through 360° about the y axis the part of the curve $y = kx^2$ which lies between the origin and the line $y = h$.

(i) Draw a diagram of the bowl and, by integration, find the volume of the solid of revolution contained within it.

(ii) Show that the centre of mass of this solid is at the point $(0, \frac{2}{3}h)$ and hence write down its distance below the top of the glass. (Notice that this is independent of the value of k).

(iii) The bowl of the glass can be modelled as a solid block of depth 9 cm with another similar block of depth 8.8 cm scooped out. Make

Exercise 4B continued

out a table similar to that on p. 130 using the results obtained in part (i). Show that the centre of mass of the bowl is 4.45 cm below the rim.

(iv) The complete wine glass can be modelled by a circular base of radius 3 cm, a cylindrical stem of height 8 cm and the bowl (as above) of depth 9 cm. The masses of these three parts are in the ratio 1:1:2. Find the distance of the centre of mass of the whole glass from the circular base.

(v) If the glass is carried on a tray, what is the maximum angle to the horizontal that the tray can be held without the glass toppling over. You may assume that the glass does not slide.

Investigations

Stability

If a solid body is made of a hemisphere with a cone of the same radius attached to its plane surface, where should its centre of mass be so that it will:

(i) always return to the upright position when knocked over?

(ii) always roll over onto its conical surface when slightly disturbed from the upright position?

Assume that the density of the body is uniform and that it is placed on a horizontal surface.

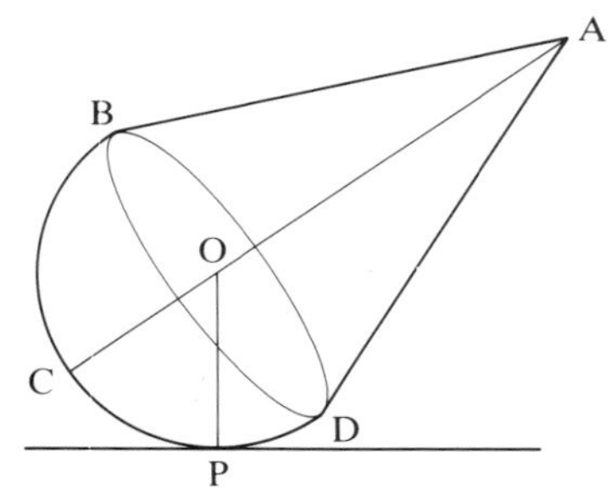

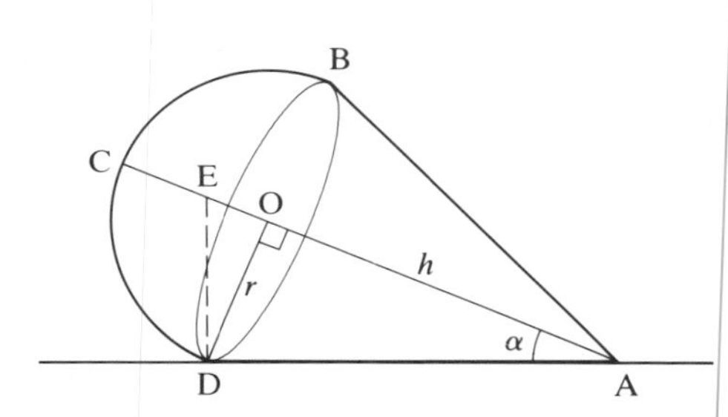

Look at the cases where the centre of mass is:

(a) between A and O;

(b) between O and E; (Hint; show that OE $= r\tan\alpha$)

(c) between E and C.

What will happen if the body is placed on a sloping surface?

Rolling

Carefully place an egg near the edge of a rectangular tray or board and gently lift this edge. Notice how the egg rolls. Most eggs will roll in a curve.

How can you make one roll in a straight line?

What can you say about the position of the egg's centre of mass when it does roll in a straight line?

Investigation continued

Investigate the manner in which different objects such as a ball, a cylinder, a coin, an apple and a cone (or part of a cone such as a plant pot) roll on a smooth surface.

What makes a body roll in a curved line rather than a straight one?

Bowls

The balls used in the game of bowls are solids of revolution which are designed to roll in curves. Contrary to popular belief they are not weighted in any way.

Investigate the design and behaviour of carpet bowls.

Centres of mass of plane regions

Activity

Find the position of the centre of mass of a semi-circular protractor by suspending it from one corner. In what ratio does it divide the axis of symmetry?

Vocabulary

A thin plane surface, like a protractor, is called a *lamina* (plural *laminae*). The laminae considered in this chapter are all uniform; that is they have the same mass per unit area, throughout. The *centroid* of a plane region is the centre of mass of a lamina of the same shape and negligible thickness.

Calculus methods for determining the centre of mass of a lamina

When a lamina or plane region is bounded by curves for which the equations are known, calculus methods can be used to find the position of its centre of mass in a similar manner to that used for volumes of revolution.

When a lamina occupies a region between a curve and the x axis, as shown in figure 4.8, it can be approximated by elementary strips of width δx and length y. One such strip, situated at the point (x, y) on the curve, is shown. The dot indicates the centre of mass of this elementary strip. Let σ (the Greek letter sigma), be the *mass per unit area* of the lamina.

The length of the strip is y and its width is δx, so its mass δM is $\sigma y \delta x$.

The coordinates of the centre of mass of the strip are $(x, \frac{y}{2})$ so $\bar{x}$ can be found using

$$(\Sigma \delta M)\bar{x} = \Sigma(\delta M x)$$

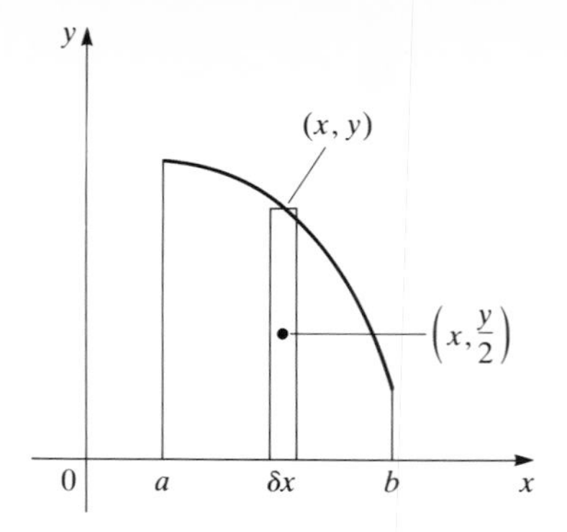

Figure 4.8

Substituting $\delta M = \sigma y \delta x$ gives

$$(\Sigma \sigma y \delta x)\bar{x} = \Sigma(\sigma y \delta x \times x)$$

If the material is uniform, the mass per unit area is constant and so the equation can be divided through by σ. It that case,

$$(\Sigma y \delta x)\bar{x} = \Sigma(yx\delta x)$$

and in the limit as $\delta x \to 0$:

$$\left(\int_a^b y\,\mathrm{d}x\right)\bar{x} = \int_a^b xy\,\mathrm{d}x \qquad ①$$

Notice that the first integral gives the area of the lamina.

You can find $\bar{y}$ in a similar way. Using the y-coordinate, $\frac{y}{2}$, of the centre of the strip and dividing by σ as before:

$$(\Sigma y \delta x)\bar{y} = \Sigma\left(y\delta x \times \frac{y}{2}\right)$$

In the limit as $\delta x \to 0$ this becomes:

$$\left(\int_a^b y\,\mathrm{d}x\right)\bar{y} = \int_a^b \frac{y^2}{2}\,\mathrm{d}x \qquad ②$$

The equations ① and ② can be combined to give the co-ordinates of the centre of mass of the lamina in form of a position vector:

$$A\begin{pmatrix}\bar{x}\\ \bar{y}\end{pmatrix} = \begin{pmatrix}\displaystyle\int_a^b xy\,\mathrm{d}x\\ \displaystyle\int_a^b \frac{y^2}{2}\,\mathrm{d}x\end{pmatrix}$$

where A is the total area of the lamina or $\int_a^b y\,\mathrm{d}x$.

EXAMPLE

Find the position of the centre of mass of a uniform semi-circular lamina of radius r.

Solution

The diagram below shows the lamina.

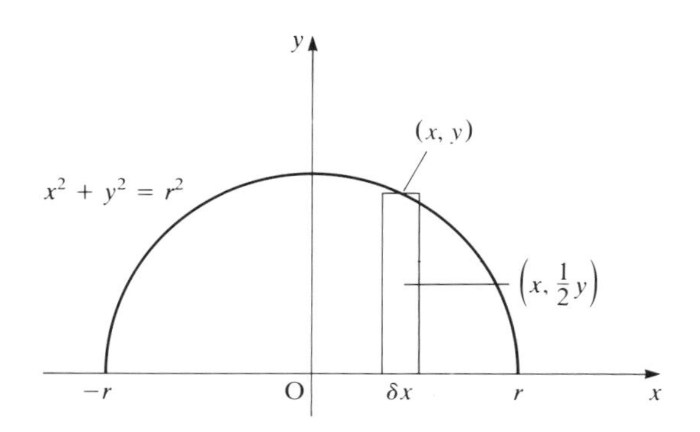

By symmetry, $\bar{x} = 0$

To find $\bar{y}$, use

$$A\bar{y} = \int_a^b \frac{y^2}{2}\,\mathrm{d}x.$$

In this case, $A = \frac{1}{2}\pi r^2$ (because the lamina is a semi-circle), and the limits are $-r$ and $+r$.

Since the equation of the curve is $x^2 + y^2 = r^2$, you can write y^2 as $r^2 - x^2$.

This gives

$$\frac{1}{2}\pi r^2\bar{y} = \int_{-r}^{r} \frac{1}{2}(r^2 - x^2)\,\mathrm{d}x$$

$$= \frac{1}{2}\left[r^2x - \frac{x^3}{3}\right]_{-r}^{r}$$

$$\Rightarrow \quad \frac{1}{2}\pi r^2\bar{y} = \frac{1}{2}\left\{\left(r^3 - \frac{r^3}{3}\right) - \left(-r^3 - \frac{(-r)^3}{3}\right)\right\}$$

$$= \frac{2}{3}r^3$$

$$\Rightarrow \quad \bar{y} = \frac{4r}{3\pi}$$

This is the distance of the centre of mass of the semi-circular lamina from its centre. It is roughly 0.4 times the radius. How does this compare with your own measurements in the Activity on page 135?

EXAMPLE

Find the co-ordinates of the centroid of the region between the curve $y = 4 - x^2$ and the positive x and y axes.

Solution

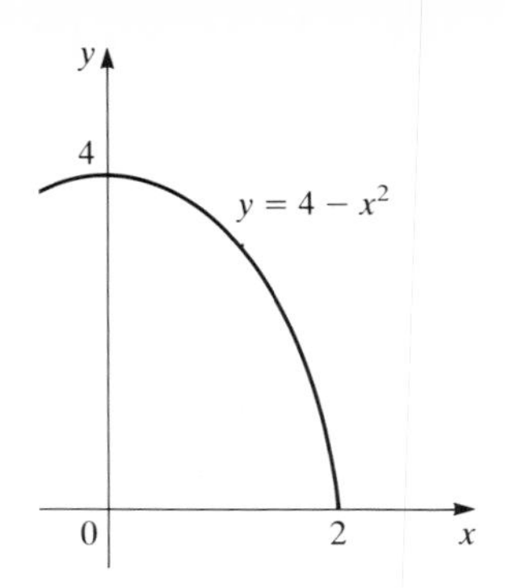

It is best to find the area, A first. In this case:

$$A = \int y\,\mathrm{d}x$$
$$= \int_0^2 (4 - x^2)\,\mathrm{d}x$$
$$= \left[4x - \frac{x^3}{3}\right]_0^2$$
$$= (8 - \tfrac{8}{3}) - 0$$
$$= \frac{16}{3}$$

Then use $A\begin{pmatrix}\bar{x}\\ \bar{y}\end{pmatrix} = \begin{pmatrix}\int_0^2 xy\,\mathrm{d}x\\ \int_0^2 \frac{y^2}{2}\,\mathrm{d}x\end{pmatrix}$ to give

$$A\bar{x} = \int_0^2 (4 - x^2)\,x\,\mathrm{d}x \qquad \text{and} \qquad A\bar{y} = \int_0^2 \frac{(4 - x^2)^2}{2}\,\mathrm{d}x$$

$$= \int_0^2 (4x - x^3)\,\mathrm{d}x \qquad\qquad = \int_0^2 \left(\frac{16 - 8x^2 + x^4}{2}\right)\mathrm{d}x$$

$$= \left[\frac{4x^2}{2} - \frac{x^4}{4}\right]_0^2 \qquad\qquad = \frac{1}{2}\left[16x - \frac{8x^3}{3} + \frac{x^5}{5}\right]_0^2$$

$$= 4 \qquad\qquad = \frac{128}{15}$$

Substituting $A = \dfrac{16}{3}$ gives

$$\bar{x} = 4 \times \frac{3}{16} \qquad \text{and} \qquad \bar{y} = \frac{128}{15} \times \frac{3}{16}$$

$$\bar{x} = \frac{3}{4} = 0.75 \qquad\qquad \bar{y} = \frac{8}{5} = 1.6$$

The coordinates of the centroid are (0.75,1.6).

Using strips parallel to the x axis

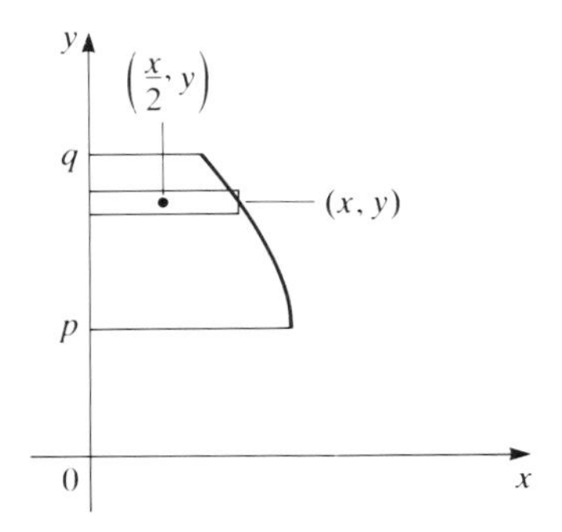

Figure 4.9

Sometimes it is necessary (or easier) to divide the region into strips parallel to the x axis as shown in figure 4.9. The area A is then $\int_p^q x\,dy$ and

$$A\begin{pmatrix}\bar{x}\\ \\ \bar{y}\end{pmatrix} = \begin{pmatrix}\displaystyle\int_p^q \frac{x^2}{2}\,dy\\ \\ \displaystyle\int_p^q yx\,dy\end{pmatrix}$$

In this case the x in each integral should be written in terms of y. Where there is a choice between methods you should choose the one which gives the easiest integral to evaluate.

Exercise 4C

1. The regions a) to d) below are all symmetrical about one axis and are defined in terms of the lines and curves which form their boundaries. For each region:
   (i) Draw a sketch of the region and, using symmetry, write down one co-ordinate of the centroid G.
   (ii) Show on your diagram an elementary strip parallel to the y axis situated at the point (x,y) on the curve and mark a dot at its centre. Show the co-ordinates of this dot on your diagram.
   (iii) Write down the area of the strip in terms of x and δx.

Exercise 4C continued

(iv) Use $A\bar{x} = \int xy\,dx$ or $A\bar{y} = \int \frac{y^2}{2}dx$, as appropriate, to find the other co-ordinate of G.

(a) The curve $y = 4 - x^2$ and the x axis.
(b) The curves $y = e^x$ and $y = -e^x$, the y axis and the line $x = 1$.
(c) The curve $y^2 = 4x$ (i.e. $y = \pm 2\sqrt{x}$) and the line $x = 4$. (Take care over the length of the strip).
(d) The curve $y = \sin x$ between $x = 0$ and $x = \pi$ and the x axis.

2. In the diagram, OAB is a uniform right-angled triangular lamina. OA = 6 cm and OB = 12 cm.

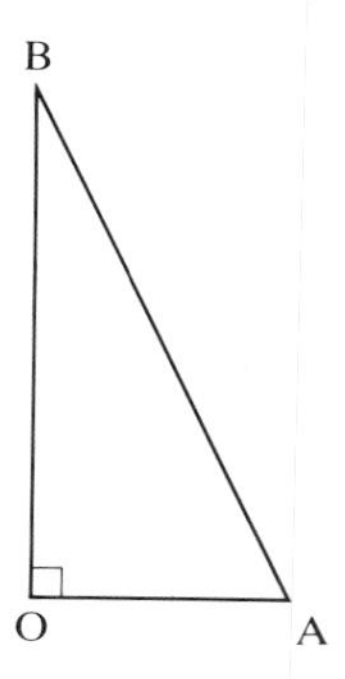

(i) Write down the area of the triangle.
(ii) Using OA as the x axis and OB as the y axis, find the equation of the line AB.
(iii) Find, by integration, the co-ordinates $(\bar{x},\bar{y})$ of the centre of mass G, of the triangle and verify that $\bar{x} = \frac{1}{3}$OA and $\bar{y} = \frac{1}{3}$OB.

3. (i) Sketch the curve $y = x^2(3 - x)$.
(ii) Find the co-ordinates of the centroid of the region between this curve and the x axis. (Note that this region is not symmetrical.)

4. (i) Draw a diagram showing the region R bounded by the curve $y^2 = 4x$ and the line $x = 4$. Mark on your diagram a strip parallel to the x axis, passing through a general point (x,y) on the curve.
(ii) Write down the length and area of the strip and the x co-ordinate of its centroid.
(iii) Find, by integrating with respect to y, the centre of mass of the region R.

In questions 5 and 6 you can use the result for the position of the centre of mass of a semi-circle found on page 137. It is at a distance $\frac{4r}{3\pi}$ from the centre.

5. A letter P is made up from a rectangle and a semi-circle with a smaller semi-circle cut out as shown.

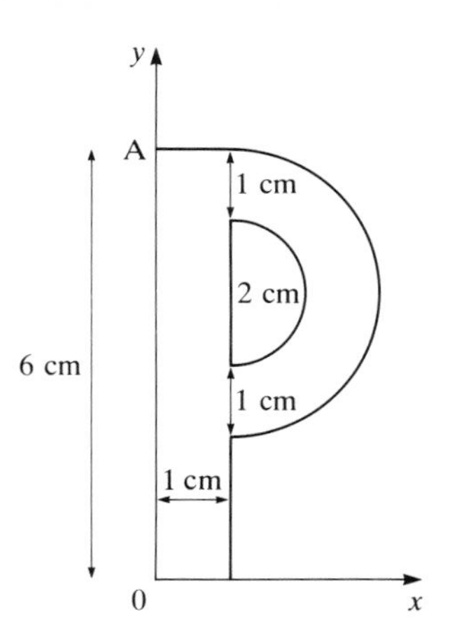

(i) Taking O as the origin, write down the areas of the component parts of the letter, and the co-ordinates of their centres of mass.

(ii) Find the co-ordinates of the centre of mass of the letter.

(iii) What angle will AO make with the vertical if the letter is hung from the corner A?

6. (i) Draw a diagram to show the circle $x^2 + y^2 = 4$ and the line $x = 1$.

(ii) Show that the area of the minor segment of the circle $x^2 + y^2 = 4$ which is cut off by the line $x = 1$ is

$$A = \frac{4\pi}{3} - \sqrt{3}$$

(iii) Find the x co-ordinate of the centre of mass of this segment by finding a suitable integral between the limits 1 and 2.

(iv) A crescent is formed by removing a segment of this shape from a semicircle of radius $\sqrt{3}$ units. Find the position of the centre of mass of this crescent.

(v) Find also the angle between the diameter of the semicircle and the vertical when the crescent is hung from one end of this diameter.

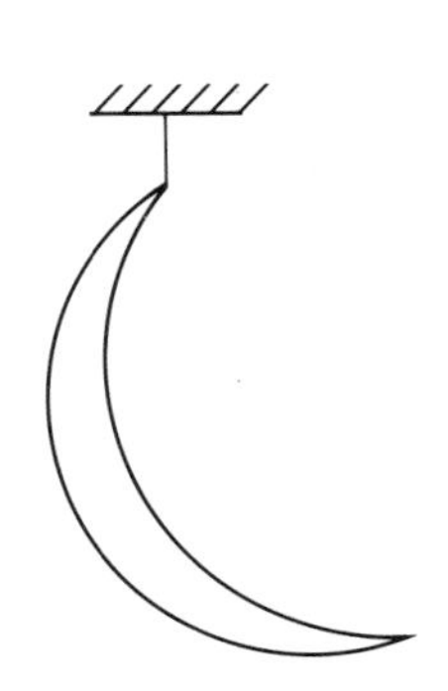

K E Y · P O I N T S

Volumes of revolution

- About the x axis:

$$V = \int_a^b \pi y^2 \, dx$$

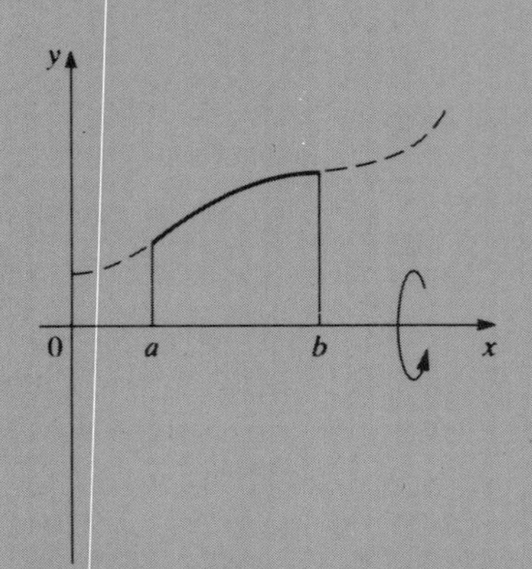

- About the y axis:

$$V = \int_p^q \pi x^2 \, dy$$

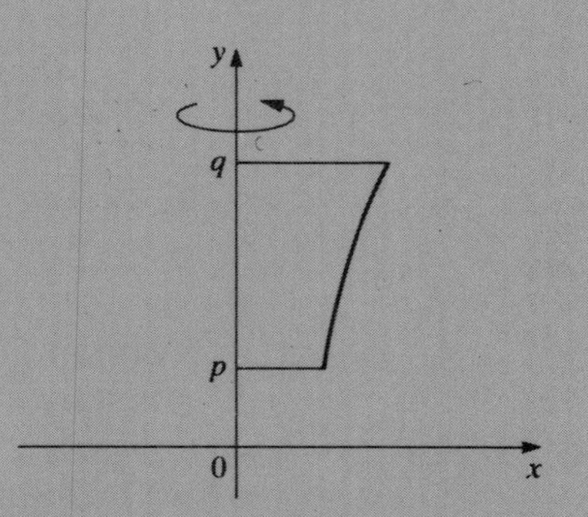

Centres of mass of uniform bodies

1 For a volume of revolution

- About the x axis:

$$\left(\int_a^b \pi y^2 \, dx\right)\bar{x} = \int_a^b \pi x y^2 \, dx$$

and $\bar{y} = 0$

- About the y axis:

$\bar{x} = 0$ and

$$\left(\int_p^q \pi x^2 \, dy\right)\bar{y} = \int_p^q \pi y x^2 \, dy$$

2 For a uniform plane lamina, area A as shown in the diagrams

- $$A\begin{pmatrix}\bar{x}\\ \bar{y}\end{pmatrix} = \begin{pmatrix}\int_a^b xy \, dx\\ \int_a^b \frac{y^2}{2} \, dx\end{pmatrix}$$

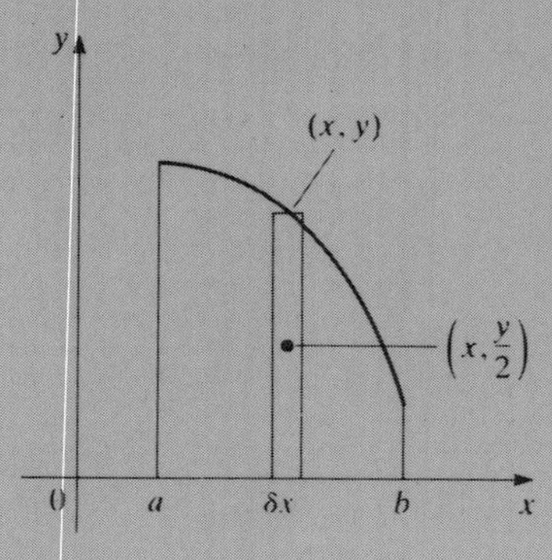

- $$A\begin{pmatrix}\bar{x}\\ \bar{y}\end{pmatrix} = \begin{pmatrix}\int_p^q \frac{x^2}{2} \, dy\\ \int_p^q yx \, dy\end{pmatrix}$$

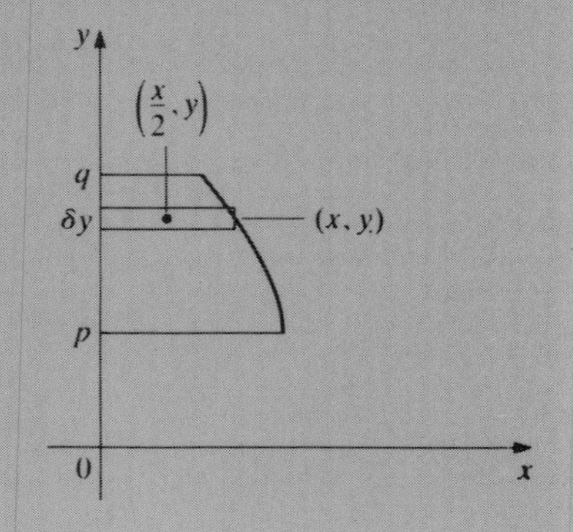

Dimensions and units

But whatever his weight in pounds, shillings and ounces,
He always seems bigger because of his bounces.

A.A. Milne

What makes the above rhyme, describing Tigger in *The House at Pooh Corner*, sound so ridiculous?

Look at the units in the first line of the rhyme. Pounds can be units of either money or mass, shillings are units of money and ounces are units of mass. You can see at once that not only is there a mixture of units, but also of the underlying quantities which they are measuring.

In this chapter you will look at the quantities which you have met in mechanics and classify them according to their *dimensions*. The dimensions of a quantity are closely related to the units in which it is measured.

All of the quantities you have met so far can be described in terms of three fundamental quantities, Mass, Length and Time. Take, for example, the quantity *area*; there are many familiar ways of finding area, depending on the shape involved (see figure 5.1).

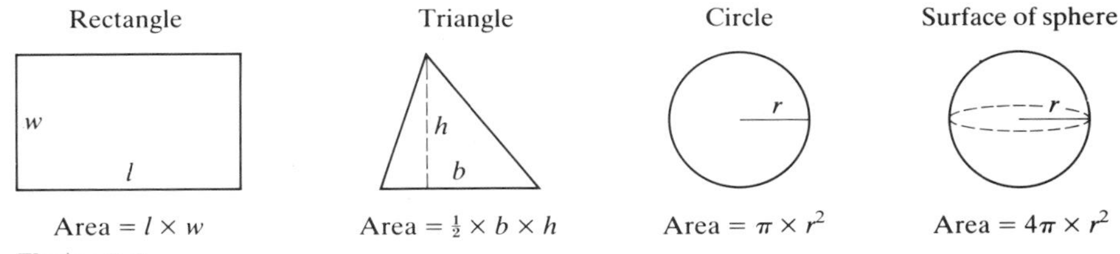

Figure 5.1

All of these formulae have essentially the same form:

$$\text{area} = \text{number} \times \text{length} \times \text{length}.$$

In the case of the rectangle the 'number' is 1 and the two 'lengths' are the length and breadth of the shape. For the circle and the sphere the two 'lengths' are the same, namely the radius, and the 'numbers' are multiples of π. However, the structure of the formula is the same for all four shapes, and indeed for any other area formula you can think of (for example, the area of a trapezium or the surface area of a cone).

The dimensions of the quantity Area can therefore be summarised using the formula

$$[\text{Area}] = \text{L}^2.$$

The square brackets [ ] mean *the dimensions of*, and the letter L represents the quantity Length. So the whole statement reads:

The dimensions of area = Length squared.

There are two important points to notice about this.

1. The numbers at the start of the formulae (in the four cases above they are 1, $\frac{1}{2}$, π and 4π) do not feature in the dimensions of the quantity because they are dimensionless.
2. The units in which a quantity is measured are derived directly from its dimensions. In the SI system the unit of length is 1 metre. Since the dimensions of area are L^2, it follows that the SI unit for area is 1 metre squared, usually written m^2.

NOTE

The area under a curve is often found by calculus methods, using the formula

$$\text{area} = \int y \, dx.$$

In this case both y and the infinitessimal quantity dx *represent lengths and so the integral has dimensions* L^2*, as must be the case since it represents an area.*

For Discussion

How many formulae can you find for the volumes of objects?

Show that all your formulae for volume involve multiplying three lengths together so that

$$[\text{Volume}] = L^3.$$

The dimensions of further quantities

So far the quantities discussed, area and volume, have only involved lengths. What are the dimensions of speed (or velocity), and of acceleration?

To find the dimensions of any quantity, start by writing down a simple formula which you might use to calculate it. For example, to calculate speed you might use

$$\text{Speed} = \frac{\text{Distance}}{\text{Time}}.$$

The dimensions follow immediately from this. A distance is clearly a length, so

$$[\text{Speed}] = \frac{[\text{Distance}]}{[\text{Time}]} = \frac{L}{T} = LT^{-1}$$

It follows that the SI unit for speed is 1 metre per second, written ms^{-1}. Notice that a second fundamental dimension, T (Time), is now involved.

Similarly $$\text{Acceleration} = \frac{\text{Change in velocity}}{\text{Time taken}}$$

and so $$[\text{Acceleration}] = \frac{[\text{Speed}]}{[\text{Time}]}$$
$$= \frac{LT^{-1}}{T}$$
$$= LT^{-2}$$

The corresponding SI unit is therefore $1\,ms^{-2}$.

Exactly the same procedure allows you to find the dimensions of force.

$$\text{Force} = \text{Mass} \times \text{Acceleration}$$
$$[\text{Force}] = [\text{Mass}] \times [\text{Acceleration}]$$
$$= M \times LT^{-2}$$
$$= MLT^{-2}.$$

Therefore the SI unit for force is 1 kilogram metre per second squared. Because force is an important concept and this unit would be such a mouthful, it is given a name of its own, the newton (N).

$$1\,N = 1\,kgms^{-2}.$$

Notice that now all three of the basic dimensions, M, L and T are being used.

EXAMPLE

(i) Find the dimensions of (a) kinetic energy, (b) gravitational potential energy and (c) work.
(ii) Comment on the significance of your answers.

Solution

(i) (a)
$$\text{Kinetic energy} = \tfrac{1}{2}mv^2$$
$$[\text{K.E.}] = M \times (LT^{-1})^2$$
$$= ML^2T^{-2}$$

(b)
$$\text{Gravitational potential energy} = mgh$$
$$[\text{P.E.}] = M \times LT^{-2} \times L$$
$$= ML^2T^{-2}$$

g is an acceleration so it has dimension LT^{-2}

(c)
$$\text{Work} = \text{Force} \times \text{Distance}$$
$$= MLT^{-2} \times L$$
$$= ML^2T^{-2}$$

(ii) The fact that all three have the same dimensions shows that all three are examples of the same underlying quantity, energy. The SI unit for this could be written as $1\,kgm^2s^{-2}$ but is actually given the special name of 1 joule, J.

Dimensionless quantities

Some quantities have no dimensions. For example, angles are dimensionless. The angle θ in figure 5.2 is defined as

$$\theta \text{ (radians)} = \frac{\text{arc length}}{\text{radius}}$$

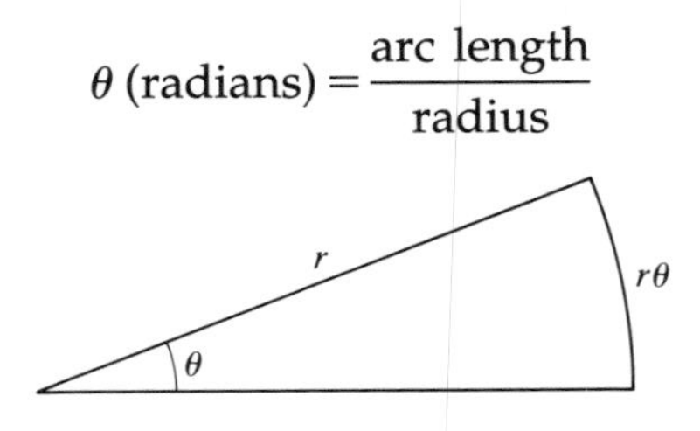

Figure 5.2

Consequently $$[\theta] = \frac{\mathrm{L}}{\mathrm{L}}$$

and θ is dimensionless.

Notice that although in this case the formula used gave θ in radians, the same result would have been obtained for θ in degrees, using

$$\theta = 360 \times \frac{\text{arc length}}{2\pi \times \text{radius}}.$$

The dimensions of a quantity cannot be altered by changing the units in which it is measured.

All numbers are dimensionless, including the trigonometrical ratios, for example $\sin\theta$, $\cos\theta$, $\tan\theta$, and irrational numbers such as π and e. This is the reason why you ignore any numbers in a formula when considering its dimensions; numbers, being dimensionless, cannot be included in this analysis.

Other systems of units

The units used in this chapter, and indeed in all the books in this series, are almost entirely SI units. There are other self-consistent sets of units, such as the 'cgs' system (centimetre, gramme, second) or the Imperial system (foot, pound, second). Knowing the dimensions of a quantity allows you to find the appropriate unit within any system.

The set of dimensions may also be extended to include quantities that arise in electricity and magnetism. For instance Q represents the dimension of electrical charge.

Change of units

It is helpful to know the dimensions of a quantity when you are changing the units in which it is to be measured, as in the following example.

EXAMPLE The energy of a body is 5000 foot poundals (imperial units based on feet, pounds and seconds). Write this in SI Units, i.e. joules. (To 3 significant figures, 1 pound (lb) = 0.453 kg and 1 foot = 0.305 m.)

Solution

The dimensions of energy are $ML^2T^{-2} = \dfrac{ML^2}{T^2}$

So
$$1 \text{ foot poundal} = \frac{1\,\text{lb} \times (1 \text{ foot})^2}{(1 \text{ second})^2}$$
$$= \frac{0.453\,\text{kg} \times (0.305\,\text{m})^2}{(1 \text{ second})^2}$$
$$= 0.0421 \text{ joules}$$

and
$$5000 \text{ foot poundals} = 5000 \times 0.0421 \text{ joules}$$
$$= 211 \text{ joules (to 3 significant figures).}$$

Exercise 5A

Draw up a table like the one below, and having copied the first two lines, extend it to cover the following quantities.

Speed, acceleration, g, force (=mass × acceleration), weight, kinetic energy, gravitational potential energy, work, power, impulse, momentum, pressure (=force/area), density, moment, angle, angular velocity, angular acceleration, modulus of elasticity, stiffness, gravitational constant (G), period, frequency, coefficient of restitution, coefficient of friction.

Quantity	Formula	Dimensions	SI unit
Area	$l \times \omega$	L^2	m^2
Volume	$l \times \omega \times h$	L^3	m^3

You may find it helpful to keep the completed table somewhere handy for reference.

Dimensional consistency

In any equation or formula, all the terms must have the same dimensions. If that is the case, the equation is said to be *dimensionally consistent*. If not, for example if force and area are being added or subtracted, or if a mass is being equated to a velocity, the equation is said to be dimensionally inconsistent.

Any statement that is dimensionally inconsistent **must** be wrong. In the rhyme at the start of this chapter, the phrase *"his weight in pounds, shillings and ounces"* is nonsense because it is dimensionally inconsistent.

EXAMPLE

Show that the formula

$$s = ut + \tfrac{1}{2}at^2$$

is dimensionally consistent.

Solution

There are three terms, with dimensions as follows:

$$[s] = \mathrm{L}$$
$$[ut] = \mathrm{LT^{-1}} \times \mathrm{T} = \mathrm{L}$$
$$\left[\frac{1}{2}at^2\right] = \mathrm{LT^{-2}} \times \mathrm{T^2} = \mathrm{L}$$

All three terms have the same dimensions, and so the formula is dimensionally consistent.

The fact that a formula is dimensionally consistent does not mean it is necessarily right, but if it is dimensionally inconsistent it is certainly wrong.

Good mathematicians develop the habit of automatically checking the dimensional consistency of anything they write. A consequence of this is that they tend to leave everything written in symbols until the final calculation. You know, for example, that g has dimensions $\mathrm{LT^{-2}}$ but if it is replaced by an approximate value, e.g. 10, its dimensions are no longer evident: $h = \frac{1}{2}gt^2$ is dimensionally consistent but $h = 5t^2$ seems not to be.

EXAMPLE

A mathematician writes down the equation

$$\frac{1}{2}m_1 v_1{}^2 = m_2 gh + m_2 v_2.$$

Show that it must be wrong.

Solution

Checking for dimensional consistency:

$$\left[\tfrac{1}{2}m_1 v_1{}^2\right] = \mathrm{M} \times (\mathrm{LT^{-1}})^2 \quad = \mathrm{ML^2T^{-2}}$$
$$[m_2 gh] = \mathrm{M} \times \mathrm{LT^{-2}} \times \mathrm{L} = \mathrm{ML^2T^{-2}}$$
$$[m_2 v_2] = \mathrm{M} \times \mathrm{LT^{-1}} \qquad = \mathrm{MLT^{-1}}.$$

The third term is dimensionally different from the other two and so the equation is incorrect.

Finding the form of a relationship

It is sometimes possible to determine the form of a relationship just by looking at the dimensions of the quantities likely to be involved, as in the following example.

A pendulum consists of a light string of length l with a bob of mass m attached to the end. You would expect that the period, t, might depend in some way on the variables l and m, and also on the value of g.

This can be expressed in the form

$$t = kl^{\alpha}m^{\beta}g^{\gamma}$$

where k is dimensionless and the powers α, β and γ are to be found.

Writing down the dimensions of each side gives

$$\begin{aligned} \mathrm{T} &= \mathrm{L}^{\alpha} \times \mathrm{M}^{\beta} \times (\mathrm{LT}^{-2})^{\gamma} \\ \mathrm{T} &= \mathrm{L}^{\alpha+\gamma}\mathrm{M}^{\beta}\mathrm{T}^{-2\gamma}. \end{aligned}$$

The left hand side of the equation, T, may be written as $\mathrm{L}^0\mathrm{M}^0\mathrm{T}^1$ and so

$$\mathrm{L}^0\mathrm{M}^0\mathrm{T}^1 \quad = \quad \mathrm{L}^{\alpha+\gamma}\mathrm{M}^{\beta}\mathrm{T}^{-2\gamma}.$$

Equating the powers of L, M and T gives:

$$\begin{aligned} &\mathrm{L}: && 0 = \alpha + \gamma \\ &\mathrm{M}: && 0 = \beta \\ &\mathrm{T}: && 1 = -2\gamma \end{aligned}$$

Solving these gives $\alpha = \frac{1}{2}$, $\beta = 0$ and $\gamma = -\frac{1}{2}$ and so the relationship is

$$t = k\sqrt{\frac{l}{g}}.$$

You may remember that you derived this formula in the chapter on simple harmonic motion on page 100. You found it in the form

$$t = 2\pi\sqrt{\frac{l}{g}}.$$

That result was obtained by writing down the equation of motion of the pendulum bob and comparing it with the equation for simple harmonic motion. In this method all you had to do was to think about the dimensions, a very beautiful piece of mathematics. Of course the dimensions method does not tell you that the value of k is 2π, but it does provide the correct form of the relationship.

Notice that the mass of the pendulum bob, m, does not feature in the formula for the period. This is found to be the case in practice: a heavier bob makes no difference to the period.

What about the angle of swing of the pendulum, θ? Since angles are dimensionless it cannot feature in this argument. You may like to think of its effect as being included within the dimensionless k. In fact the value of

k is approximately constant (equal to 2π) for small values of θ, but, it does vary with θ for larger swings.

In the next example the work is taken a step further. First the form of a relationship is proposed, then data are used to determine the constant involved. The resulting formula is used to predict the outcome of a future experiment. Notice that the units look after themselves providing you are consistent with their use; in this case they happen to be in the cgs system.

EXAMPLE

In an experiment, small spheres are dropped into a container of liquid which is sufficiently deep for them to attain terminal velocity. For any sphere, the terminal velocity, V, is thought to depend on its radius, r, its weight mg and the viscosity of the liquid, η (the Greek letter 'eta'). (Viscosity is a measure of the 'stickiness' of a liquid and has dimensions $ML^{-1}T^{-1}$.)

(i) Write down a formula for V as the product of unknown powers of r, mg, η, and a dimensionless constant k.

(ii) Find the powers of r, mg and η and write out the formula for V with these values substituted in.

In a particular liquid it is found that a sphere of mass $0.02\,g$ and radius $0.1\,cm$ has terminal velocity $5\,cm s^{-1}$.

(iii) Find the value of $\frac{k}{\eta}$ taking g to be $1000\,cm s^{-2}$.

(iv) Find the terminal velocity of a sphere of mass $0.03\,g$ and radius $0.2\,cm$.

Solution

(i) $V = kr^{\alpha}(mg)^{\beta}\eta^{\gamma}$

(ii) Taking dimensions of both sides of the equation gives

$$LT^{-1} = L^{\alpha} \times (MLT^{-2})^{\beta} \times (ML^{-1}T^{-1})^{\gamma}$$
$$M^0L^1T^{-1} = M^{\beta+\gamma}L^{\alpha+\beta-\gamma}T^{-2\beta-\gamma}$$

Equating powers:
M: $0 = \beta + \gamma$
L: $1 = \alpha + \beta - \gamma$
T: $-1 = -2\beta - \gamma$

Solving these gives $\alpha = -1$, $\beta = 1$ and $\gamma = -1$

and so the formula is $$V = \frac{kmg}{r\eta}$$

(iii) Substituting $m = 0.02$, $r = 0.1$, $g = 1000$ and $V = 5$ gives

$$5 = \frac{k \times 0.02 \times 1000}{0.1 \times \eta}$$

$$\frac{k}{\eta} = 0.025 \quad \text{(in cgs units).}$$

(iv) Substituting $m = 0.03$, $r = 0.2$, $g = 1000$ and $\frac{k}{\eta} = 0.025$ gives

$$V = \frac{0.025 \times 0.03 \times 1000}{0.2}$$

$$= 3.75$$

The terminal velocity of this sphere is $3.75\,\text{cms}^{-1}$.

The method of dimensions

This method for finding the form of a proposed formula is sometimes called *the method of dimensions*; at other times it is simply referred to as *dimensional analysis*. There are a number of points which you should realise when using it.

1. In mechanics relationships are based on three fundamental quantities Mass, Length and Time and so the right hand side of the formula can involve no more than three independent quantities. Otherwise you will end up with three equations to find four or more unknown powers. (The last example actually involved four quantities, m, g, r and η, but the first two of these were tied together as the weight, mg, and were not, therefore, independent.)
2. The method requires you to make modelling assumptions about which quantities are going to be important and which can be ignored. You must always be prepared to review these assumptions.
3. This method can only be used when a quantity can be written as a product of powers of other quantities. There are many situations which cannot be modelled by this type of formula and for these situations this method is not appropriate. For example, the formula $s = ut + \frac{1}{2}at^2$ has two separate terms which are added together on the right hand side. The form of this relationship could not be predicted using the method of dimensions.

Exercise 5B

1. Check the dimensional consistency of each of the formulae below and hence state whether it could be correct or is definitely wrong. You may assume that all letters have their conventional meanings.
   (a) $v^2 - u^2 = 2as$
   (b) $x = \dfrac{(m_1x_1 + m_2x_2)}{m_1 + m_2}$

Exercise 5B continued

(c) $F - mg = ma$

(d) $T_1 - T_2 = \frac{1}{2}mv^2$ (T_1 and T_2 are tensions)

(e) $Fs = mv - mu$

(f) $mg\sin\alpha - \mu mg\cos\alpha = ma$

(g) $m_1gd_1 - m_2gd_2 = (m_1 + m_2)a$

(h) $\frac{1}{2}mv^2 = \frac{1}{2}mu^2 + mgh\sin\theta$

(i) $v^2 = \omega^2(a^2 - x^2)$

(j) $\frac{1}{2}F(u + v) = \dfrac{(v^2 - u^2)}{2t}$

2. A scientist thinks that the speed, v, of a wave travelling along the surface of an ocean depends on the depth of the ocean, h, the density of the water in the ocean, ρ (the Greek letter 'rho') and g.

(i) Write down the dimensions of each of the quantities v, h, ρ and g.

The scientist expresses his idea using the formula

$$v = kh^\alpha \rho^\beta g^\gamma \qquad \text{where } k \text{ is dimensionless.}$$

(ii) Use dimensional analysis to find the values of α, β and γ and write out the formula with these values substituted in.

(iii) Do waves travel faster in deep water or shallow water?

(iv) Do waves travel faster in winter (when the water density is greater) or in summer?

3. A water container has a hole in the bottom. When it is filled to a depth d it takes time t for the water to run out. The time t may be modelled as the product of powers of d and g and a dimensionless constant k.

(i) Write down the formula for t.

(ii) Use dimensional analysis to find the powers of d and g in the formula.

When the container is filled to a depth of $0.4\,\text{m}$ it takes 30 seconds to empty.

(iii) Taking g to be $10\,\text{ms}^{-2}$, find the value of k.

(iv) How long does it take the container to empty when it is filled to a depth of $60\,\text{cm}$?

4. (i) Write down the dimensions of energy.

A student who has not yet learnt about the energy of rotating bodies observes a symmetrical flywheel of mass m and radius r rotating in a horizontal plane about a vertical axis through its middle with angular speed ω. The student decides that the rotational energy, E, of the flywheel can only depend on a product of powers of these three quantities and a dimensionless constant.

(ii) Express the student's opinion in the form of a formula for E.

(iii) Use dimensional analysis to find the powers of m, r and ω in the formula.

The student conducts some experiments which show that when $m = 20\,\text{kg}$, $r = 0.5\,\text{m}$ and $\omega = 4$ radians per second, $E = 40$ joules.

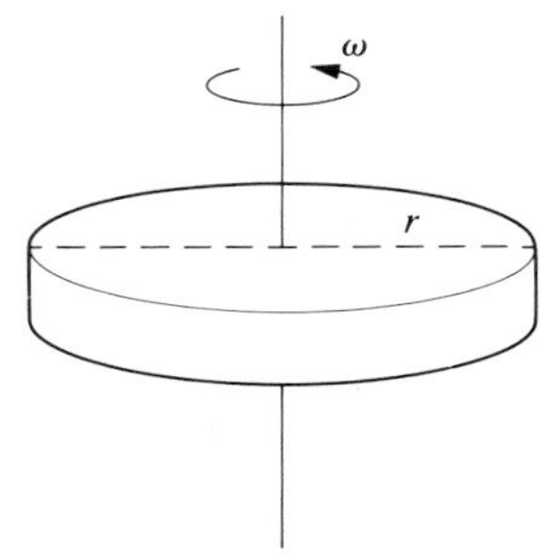

(iv) Given that the student's experiments are accurate, find the complete formula for E.

(v) A similar flywheel, mounted in the same way, has mass $50\,\text{kg}$ and radius $1.2\,\text{m}$. Find its energy when it is rotating at 12 radians per second.

(vi) Suggest how the student's experiments might have been carried out.

5. A ball of mass m travelling with velocity v vertically downwards hits a horizontal surface and bounces back up, rising to a height h. The coefficient of restitution between the ball and the surface is e.

(i) Use the definition of coefficient of restitution to show that e is dimensionless.

It is believed that h depends on a product of powers of v, m and g and a dimensionless constant, so a formula is proposed of the form

$$h = kv^{\alpha}m^{\beta}g^{\gamma} \qquad \text{where } k \text{ is dimensionless.}$$

(ii) Use dimensional analysis to find the values of α, β and γ, and write down the formula with these values substituted in.

(iii) Two balls, identical in all respects except that one is heavier than the other, fall with equal speeds. Explain how your answer to part (ii) allows you to decide which (if either) bounces higher.

(iv) The height h clearly does depend on e also. Why was it not considered useful to include it in the formula?

(v) Use your knowledge of mechanics to find a formula for h in terms of v, m, g and e.

(vi) Explain the relationship between k in the answer to part (ii) and e.

6. The magnitude of the force of gravitational attraction, F, between two objects of mass m_1 and m_2 at a distance d apart is given by

$$F = G\frac{m_1 m_2}{d^2}$$

where G is the universal constant of gravitation.

(i) Find the dimensions of G.

An astronomer proposes a model in which the lifetime, t, of a star depends on a product of powers of its mass, m, its initial radius r_0, G and a dimensionless constant.

Exercise 5B continued

(ii) Use the method of dimensions to find the resulting formula for t.

Observation shows that the larger the initial radius the longer the lifetime of the star, but that the larger the mass the shorter the lifetime of the star.

(iii) Is the model consistent with these observations?

(iv) Show that the model can be expressed more simply if the initial density, ρ_0, of the star is used as one of the variables.

7. The volume rate of flow R (m^3s^{-1}) of a liquid with a viscosity η ($kg m^{-1}s^{-1}$) through a cylindrical pipe of length a (m) and an internal radius r (m) is believed to be of the form

$$R = k\eta^w a^x r^y P^z,$$

where k is a non-dimensional constant and P (Nm^{-2}) is the pressure difference between the ends of the pipe.

Using dimensional considerations

(i) show that $w = -1$ and $z = 1$;

(ii) find the relationship between x and y.

Observations of a water supply provided at constant pressure and viscosity produced the following recordings:

Pipe length, a (m)	100	200
Internal radius, r (m)	0.050	0.050
Flow rate, R (m^3s^{-1})	0.420	0.210

(iii) Derive a formula for R in terms of a and r for such a supply.

(iv) Find the rate of flow of water through 1000 m of pipe with an internal radius of 0.075 m under the same pressure and with the same viscosity. [MEI]

8. In an investigation into viscosity small spheres are allowed to fall from rest through a column of a viscous liquid and their terminal speeds are measured. It is believed that the terminal speed V of a sphere depends upon a product of powers of its weight mg, its radius r and the viscosity η of the liquid and a dimensionless constant.

(i) Given that the dimensions of viscosity are $ML^{-1}T^{-1}$ use the method of dimensions to establish the form of the relationship for V in terms of m, r, η and g.

(ii) The mass m of the sphere may be written in terms of its radius r and its density ρ. Write down this relationship.

(iii) Using (i) and (ii) write down a relationship for V in terms of ρ, r, η and g.

(iv) A steel sphere of radius 0.15 cm and density 7.8 $g cm^{-3}$ falls through a column of a viscous liquid and is found to have a terminal speed of 6.0 $cm s^{-1}$. A lead sphere of density 11.4 $g cm^{-3}$ falling through the same liquid is found to have a terminal speed of 5.0 $cm s^{-1}$. Determine the radius of the lead sphere. [MEI]

9. (i) Define π and hence find its dimensions.

(ii) Newton's Law of Universal Gravitation states that the magnitude of the force F on a point mass m_1 due to the presence of a point mass m_2 at a distance r is given by

$$F = G\frac{m_1 m_2}{r^2}$$

Find the dimensions of G.

(iii) The constant G has a value 6.67×10^{-11} in S.I. units. An astronomer proposes a system of units based upon the mass of the Earth, M_E, its radius, R_E, and the time it takes for one revolution about the Sun, T_E. Given that $M_E = 6.98 \times 10^{24}\,\text{kg}$, $R_E = 6.37 \times 10^6\,\text{m}$ and $T_E = 3.16 \times 10^7\,\text{s}$ find the value of G in terms of these Earth units.

(iv) Comment briefly upon the different natures of the constants π and G. [MEI]

10. In the early 17th century Mersenne (1588–1648) conducted experiments with long lengths of rope and so obtained the law for the frequency of transverse vibrations of strings.

Assuming that the frequency depends on products of powers of T, the tension in the rope, l, the length of the rope, and m, the mass per unit length of the rope,

(i) find, by dimensional analysis, the form of the relationship.

A rope of length $24\,\text{m}$ and mass $0.5\,\text{kg}$ per metre under tension of $72\,\text{N}$ is found to vibrate with a frequency of $\frac{1}{4}$ of a cycle per second.

(ii) State the exact relationship between the frequency, T, l and m.

(iii) Find the frequency of vibration of a string of length $20\,\text{cm}$ and mass $0.005\,\text{g}\,\text{cm}^{-1}$ under a tension of 8×10^5 dynes. (The dyne is the c.g.s. unit of force: 1 dyne is the force required to give a mass of $1\,\text{g}$ an acceleration of $1\,\text{cm}\,\text{s}^{-2}$).

11. An artillery officer is conducting a series of experiments to find the maximum range, R, of a gun which fires shells from ground level across level ground.

(i) Show, by considering the shell as a projectile with initial velocity u at an angle α to the horizontal and ignoring air resistance, that

$$R = \frac{u^2}{g}\sin 2\alpha$$

and that this has a maximum value, R_M, when α is $45°$.

(ii) Show that the expression for R in part (i) is dimensionally consistent.

The officer finds that the shells fall some distance short of their predicted range. Deciding that this must be due to air resistance, he proposes a model in which the magnitude of the force of resistance, F, is proportional to the speed of the shell, v:

$$F = cv.$$

(iii) Show that the constant c is not dimensionless and write down its dimensions.

Exercise 5B continued

The officer then suggests that a better model for the range of the shell, when fired at 45° to the horizontal, is given by:

$$R = R_M(1 - \varepsilon)$$

where ε (epsilon) is a number less than 1 representing the proportion of the range which is lost.

(iv) State the dimensions of ε.

The officer believes that ε depends on a product of powers of the speed of projection u, the constant c,g, and the mass m of the shell.

(v) Use dimensional analysis to find an expression for ε in which the power of u is 1. Express the improved model as a formula for R.

It is suggested that a more accurate and more general model still would have the form

$$R = R_M(1 + a_1\varepsilon + a_2\varepsilon^2 + a_3\varepsilon^3 + \ldots)$$

where $a_1, a_2, a_3, \ldots$ are dimensionless coefficients, taking into account (among other things) the angle of projection. (Clearly a_1 would be negative.)

(vi) Show that such a model is dimensionally consistent.

Investigations

Fundamental quantities

The system of dimensions you have met in this chapter describes the quantities relevant to mechanics on a human scale in a consistent and altogether satisfactory manner. When, however, these concepts are applied at the scale of the whole universe, difficulties arise; cosmology is a set of unsolved problems. Is it possible that we are looking at the universe in terms of the wrong basic quantities?

It is suggested that three more appropriate quantities might be: Mass, Energy and Momentum, corresponding to the Laws of Conservation of Mass, Conservation of Energy and Conservation of Momentum.

(i) Show that such a system would be inadequate.

(ii) Suggest and investigate an alternative system in which one of the quantities is replaced by another which might be considered fundamental.

Definition of force

The system of dimensions worked out in this chapter depends on the assumption that force is defined by the equation $F = ma$. An alternative definition is given by Newton's law of gravitation.

$$F = G\frac{m_1 m_2}{d^2}.$$

Suppose that instead this law is taken as the definition of force, so that in suitable units,

$$F = \frac{m_1 m_2}{d^2},$$

Investigations continued

and Newton's 2nd Law is regarded as secondary to this, so that it is written

$$F = kma.$$

What would be the dimensions of k, using the M, L, T system?

Investigate the effect of this change of definition on the dimensions of some other quantities commonly used in mechanics. Do you consider such a change to be desirable?

KEY POINTS

- Any quantity in Mechanics may be expressed in terms of the three fundamental dimensions:

  Mass, M; Length, L; Time, T.
- The unit for any quantity is derived from the three fundamental dimensions.
- Numbers are dimensionless.
- All formulae and equations must be dimensionally consistent.
- Using dimensional analysis you can sometimes find the form of a relationship as the product of powers of the quantities involved, and a dimensionless constant.

Mathematical Notes

1. Velocity and acceleration for motion in a circle

When a particle moves in a circle, it is convenient for the velocity and acceleration to be expressed in the radial and transverse (tangential) directions. For this reason, unit vectors $\hat{\mathbf{r}}$ and $\hat{\boldsymbol{\theta}}$ are defined in these two directions. Unlike the unit vectors **i** and **j**, however, these vectors are not constant. They change in direction as the particle moves round the circle.

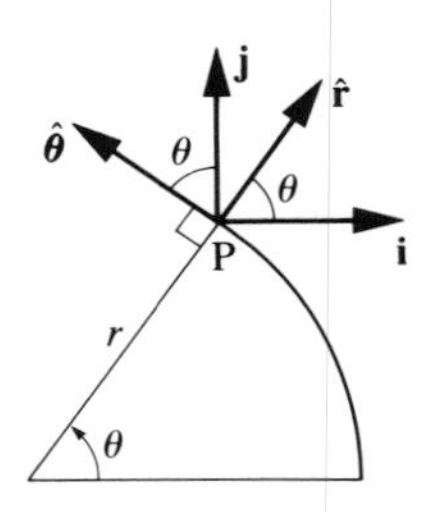

Figure A1

Figure A1 shows $\hat{\mathbf{r}}$ and $\hat{\boldsymbol{\theta}}$ when the particle P is at a general position. You can see that

$$\hat{\mathbf{r}} = \cos\theta\,\mathbf{i} + \sin\theta\,\mathbf{j}$$

and

$$\hat{\boldsymbol{\theta}} = -\sin\theta\,\mathbf{i} + \cos\theta\,\mathbf{j}$$

Differentiating $\hat{\mathbf{r}}$ with respect to time,

$$\frac{\mathrm{d}\hat{\mathbf{r}}}{\mathrm{d}t} = \frac{\mathrm{d}}{\mathrm{d}t}(\cos\theta)\mathbf{i} + \frac{\mathrm{d}}{\mathrm{d}t}(\sin\theta)\mathbf{j}$$

$$= -\sin\theta\,\frac{\mathrm{d}\theta}{\mathrm{d}t}\mathbf{i} + \cos\theta\,\frac{\mathrm{d}\theta}{\mathrm{d}t}\mathbf{j}$$

$$= (-\sin\theta\;\mathbf{i} + \cos\theta\;\mathbf{j})\dot{\theta} \qquad \text{where } \dot{\theta} = \frac{\mathrm{d}\theta}{\mathrm{d}t}.$$

You will recognise that the vector in brackets, $(-\sin\theta\ \mathbf{i}+\cos\theta\ \mathbf{j})$, is the same as $\hat{\boldsymbol{\theta}}$, the unit vector in the transverse direction. Consequently,

$$\frac{\mathrm{d}\hat{\mathbf{r}}}{\mathrm{d}t}=\dot{\theta}\hat{\boldsymbol{\theta}}.$$

Similarly,

$$\frac{\mathrm{d}\hat{\boldsymbol{\theta}}}{\mathrm{d}t}=-(\cos\theta)\dot{\theta}\ \mathbf{i}-(\sin\theta)\dot{\theta}\ \mathbf{j}$$

$$=(-\cos\theta\,\mathbf{i}+\sin\theta\,\mathbf{j})\dot{\theta}$$

i.e.

$$\frac{\mathrm{d}\hat{\boldsymbol{\theta}}}{\mathrm{d}t}=-\dot{\theta}\hat{\mathbf{r}}.$$

These two results, $\frac{\mathrm{d}\hat{\mathbf{r}}}{\mathrm{d}t}=\dot{\theta}\hat{\boldsymbol{\theta}}$ and $\frac{\mathrm{d}\hat{\boldsymbol{\theta}}}{\mathrm{d}t}=-\dot{\theta}\hat{\mathbf{r}}$ are very important, and apply to all motion described in polar co-ordinates, not just circular motion.

In the case of circular motion with radius r, the displacement from the centre at any time is $r\,\hat{\mathbf{r}}$.

Differentiating this with respect to time, you obtain

$$\begin{aligned}\text{velocity} &= \frac{\mathrm{d}r}{\mathrm{d}t}\hat{\mathbf{r}}+r\frac{\mathrm{d}\hat{\mathbf{r}}}{\mathrm{d}t}\\ &= 0+r\dot{\theta}\hat{\boldsymbol{\theta}} \qquad \text{(since } r \text{ is constant)}\\ &= r\dot{\theta}\hat{\boldsymbol{\theta}} \qquad \text{(i.e. } r\dot{\theta} \text{ in the transverse direction).}\end{aligned}$$

Differentiating again you obtain

$$\begin{aligned}\text{acceleration} &= \frac{\mathrm{d}r}{\mathrm{d}t}\dot{\theta}\hat{\boldsymbol{\theta}}+r\frac{\mathrm{d}\dot{\theta}}{\mathrm{d}t}\hat{\boldsymbol{\theta}}+r\dot{\theta}\frac{\mathrm{d}\hat{\boldsymbol{\theta}}}{\mathrm{d}t}\\ &= 0+r\ddot{\theta}\hat{\boldsymbol{\theta}}-r\dot{\theta}^2\hat{\mathbf{r}}\end{aligned}$$

Thus the acceleration has two components:

transverse: $r\ddot{\theta}$

radial: $-r\dot{\theta}^2$.

2. Solving the differential equation for SHM

Simple harmonic motion has been defined by the differential equation

$$\ddot{x}=-\omega^2x.$$

Several forms of the solution of this equation, involving sine and cosine functions, have been used in chapter 3. The work which follows shows you how this equation is solved to obtain the general solution in the form

$x = a\sin(\omega t + \varepsilon)$. You have seen in chapter 3 that other forms of the solution are equivalent to this.

The acceleration, $\ddot{x} = \frac{dv}{dt}$, can be written using the chain rule as

$$\frac{dv}{dx} \times \frac{dx}{dt} = v\frac{dv}{dx}$$

So the SHM equation becomes

$$v\frac{dv}{dx} = -\omega^2 x$$

Separating the variables gives:

$$\int v\,dv = \int -\omega^2 x\,dx$$

$$\Rightarrow \quad \frac{v^2}{2} = -\frac{\omega^2 x^2}{2} + C$$

Let the amplitude be a, so that $v = 0$ when $x = a$.

Then $$0 = -\frac{\omega^2 a^2}{2} + C$$

$$C = +\frac{\omega^2 a^2}{2}$$

Therefore $$v^2 = \omega^2(a^2 - x^2)$$

It follows from this that: $$v = \pm\omega\sqrt{(a^2 - x^2)}$$

$$\Rightarrow \quad \frac{dx}{dt} = \pm\omega\sqrt{(a^2 - x^2)}$$

Separating the variables gives: $$\int \frac{dx}{\pm\sqrt{(a^2 - x^2)}} = \int \omega\ dt$$

For the positive square root, the integral on the left hand side is $\arcsin\left(\frac{x}{a}\right)$

In this case:

$\arcsin\left(\frac{x}{a}\right) = \omega t + \varepsilon$ where ε is the constant of integration

$$\Rightarrow \quad x = a\sin(\omega t + \varepsilon) \qquad ①$$

For the negative root the left hand integral is $\arccos\left(\frac{x}{a}\right)$ and in this case:

$$\arccos\left(\frac{x}{a}\right) = \omega t + \varepsilon_1$$

$$\Rightarrow \quad x = a\cos(\omega t + \varepsilon_1) \qquad ②$$

As you saw in chapter 3, the two forms (① and ②) are equivalent.

3. A method for finding the position of the centre of mass of a shell.

It is possible to find the position of the centres of mass of thin shells by using formulae already obtained for solid bodies.

A hemispherical shell, for example, can be formed by removing a small hemisphere (say H_1) from a larger one (H_2). Suppose in the first instance that the radii of the two hemispheres are a and b respectively. The table shows the volumes, masses and positions of the centres of mass of H_1, H_2, and the shell.

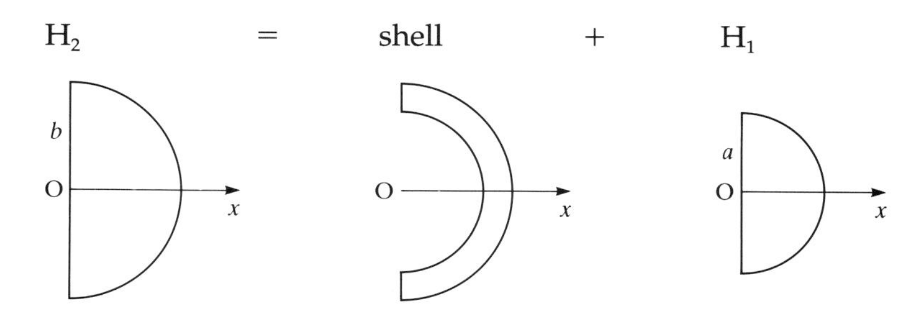

Mass	$\frac{2}{3}\pi\rho b^3$	$\frac{2}{3}\pi\rho(b^3 - a^3)$	$\frac{2}{3}\pi\rho a^3$
Distance of c.o.m. from O	$\frac{3}{8}b$	$\bar{x}$	$\frac{3}{8}a$

This gives the equation: $\frac{2}{3}\pi\rho b^3(\frac{3}{8}b) = \frac{2}{3}\pi\rho(b^3 - a^3)\bar{x} + \frac{2}{3}\pi\rho a^3(\frac{3}{8}a)$

$$(b^3 - a^3)\bar{x} = b^3(\tfrac{3}{8}b) - a^3(\tfrac{3}{8}a)$$

Therefore $(b^3 - a^3)\bar{x} = \frac{3}{8}(b^4 - a^4)$

$$\Rightarrow \quad \bar{x} = \frac{3(b^4 - a^4)}{8(b^3 - a^3)}$$

This is the position of the centre of mass of a shell of any thickness. The position of the centre of mass of a thin shell of radius a, can be found by letting $b = a + h$, and then letting $h \to 0$.

It is necessary to expand $(a+h)^4$ and $(a+h)^3$ using the binomial theorem:

Then $b^4 - a^4 = (a+h)^4 - a^4$
$= (a^4 + 4a^3h + 6a^2h^2 + 4ah^3 + h^4) - a^4$
$= 4a^3h + 6a^2h^2 + 4ah^3 + h^4$
$= h(4a^3 + 6a^2h + 4ah^2 + h^3)$

and similarly:

$$b^3 - a^3 = (a+h)^3 - a^3$$
$$= h(3a^2 + 3ah + h^2)$$

Since $$\bar{x} = \frac{3(b^4 - a^4)}{8(b^3 - a^3)}$$

We have $$\bar{x} = \frac{3h(4a^3 + 6a^2h + 4ah^2 + h^3)}{8h(3a^2 + 3ah + h^2)}$$

Cancelling h and then letting $h \to 0$ gives the distance of the centre of mass of the shell from the centre of the shell as

$$\frac{3}{8} \times \frac{4a^3}{3a^2} = \frac{a}{2}.$$

Answers

Exercise 1A

1. (i) 8.2 $rads^{-1}$ (ii) 4.7 $rads^{-1}$ (iii) 3.5 $rads^{-1}$
2. (i) 20 r.p.m. (ii) 2.1 $rads^{-1}$
3. 2865 r.p.m.
4. (i) 0.52 ms^{-1} (ii) 0.44 ms^{-1}
5. 32.5 $rads^{-1}$
6. (i) 50 $rads^{-1}$ (ii) 150 $rads^{-1}$
7. (i) 1.99×10^{-7} (ii) 7.27×10^{-5} (iii) 465.3 ms^{-1} (iv) about 290 ms^{-1}
8. 2.29:1
9. (i) 61.69 joules (ii) points on a large object would travel with different speeds
10. (i) big: 2.09×10^{-3} ms^{-1} small: 1.16×10^{-4} ms^{-1} (ii) 18:1 the radius is also involved
11. (i) 4.91 m (ii) 12.33 ms^{-1}
12. (i) $v = 2\cos t\,\mathbf{i} - 2\sin t\,\mathbf{j} - \frac{1}{2}\mathbf{k}$ (ii) $|v| = \sqrt{(4\frac{1}{4})} = 2.06$ ms^{-1} (iii) $\mathbf{a} = -2\sin t\,\mathbf{i} - 2\cos t\,\mathbf{j}$; magnitude 2 ms^{-2} horizontally towards the vertical axis
13. $\sqrt{\dfrac{5000\pi}{t+144.5\pi}}$; assuming radius increases steadily, not in steps with each wrap round.
14. (i) no

Exercise 1B

1. (i) neither (ii) no (iii) 13.5 ms^{-2}, 11.25 ms^{-2} (iv) towards the centre
2. (i) (a) neither slips (b) B slips, A doesn't (c) both slip (ii) A slips first, radius matters, mass doesn't matter
3. (i) accelerates towards centre (ii) 11.25 ms^{-2} (iii) 2250 N (iv) no
4. (i) T (ii) F (iii) F
5. (i) (a) 0.5 $rads^{-1}$ (b) 1 ms^{-2} (c) 60N towards centre (ii) skater is particle
6. B has greater force because greater acceleration
7. (i) (a) $\dfrac{(2\times10^7)}{\sqrt{r}}$ (3sf) (b) $\pi r^{\frac{3}{2}}\times10^{-7}$ s (ii) $T^2 = \pi^2 10^{-14} r^3$ (iii) 4.2×10^7 m
8. (i) 5.708×10^{-3} ms^{-2} (ii) 1.77×10^{30} kg (iii) any planet in this orbit would have the same period whatever its mass
9. (i) vertical force required to balance weight
   (ii) 12.57 $rads^{-1}$
   (iii)

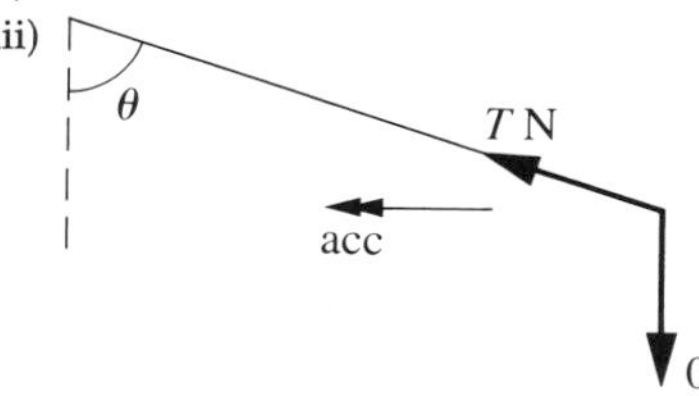

   (iv) $T\sin\theta = mrw^2 = 0.18\times0.8\sin\theta\times12.57^2$; $T\cos\theta = mg$
   (vi) 22.8 N
10. (i) 263 N (ii) 23 ms^{-1}
11. (i) 6272 N
   (ii) 69 m.p.h.
   (iii)

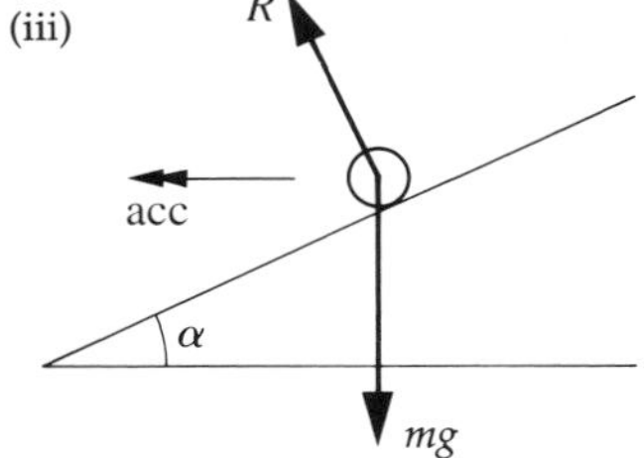

   (iv) $R\sin\alpha = \dfrac{mv^2}{r}$, $R\cos\alpha = mg$
   (vi) 10.8° or 0.189 rads
12. (i) 44 m.p.h. (ii) 4 m.p.h. faster
13. (i) π $rads^{-1}$, 5π ms^{-1} (iii) 40.5 r.p.m. (iv) 100 m

Exercise 1C

1. (i) 0.02 rads^{-2} (ii) 900 radians 2. (i) 6.28 rads^{-2} (ii)
3. (i) 4 rads^{-1} (ii) 0.0005 rads^{-2} (iii) 16 000 rad (iv) 32 000 m
4. 9.42 ms^{-1}
5. (i)

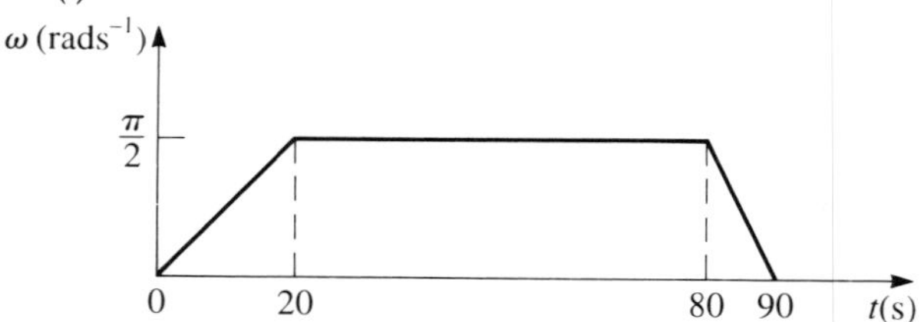

(ii) 18.75 revolutions, 353 m
(iii) 7.40 ms^{-2} centre
(iv) $\frac{3\pi}{40}$ ms^{-2} vertically upwards and $\frac{3\pi^2}{16}$ ms^{-2} horizontally towards the centre

6. (i) (a) Y (b) N (ii) (a) N (b) Y (iii) (a) N (b) Y (iv) (a) Y (b) Y
7. **A** (i) 18 m J (ii) stops; 1.84 m (iii) (a) 18 m N (b) 9 m N
   B (i) 4.75 m J (ii) stops; 0.48 m (iii) (a) 6.49 m N (b) 9.48 m N
   C (i) 22.19 m J (ii) complete revolutions; 4.98 ms^{-1} (iii) (a) 73.8 m N in (b) 15 m N in.
8. (i) 0.186 mg J (iv) 0.598 m (v) No
9. (i) 6.35 ms^{-1} (ii) $\sqrt{(9 + 1.6g(1 - \cos\theta))}$ (iii) 0.725 N, 30.125 N (iv) $T = 15.43 - 14.7\cos\theta$, $g\sin\theta$
10. (i) 82 ms^{-1}, 53 ms^{-1}
    (ii) 67.8 ms^{-2}, 28.6 ms^{-2}
    (iii) 5432 N, 1316 N

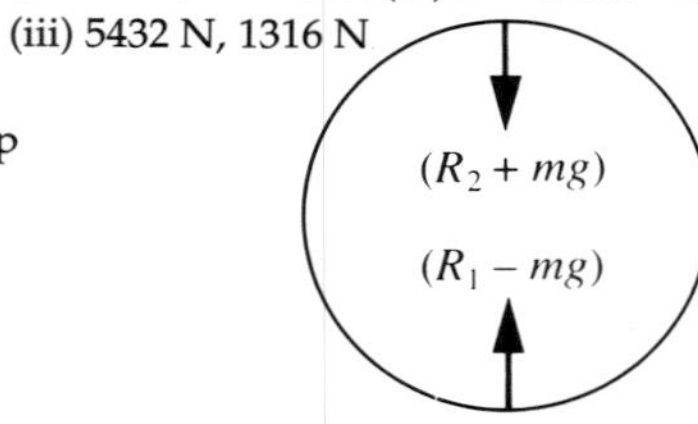

    (iv) He would fail to complete the loop
    (v) 138 m
11. (i) 15 cm (ii) 3.62 rads^{-1} (iii) about 2 rads^{-2} (iv) Girl lets go before reaching the vertical.
12. (i) almost zero speed at the top of the loop not sufficient (ii) 20 m (iii) 8 m
13. (i) 12 ms^{-1}, 27 m.p.h.

Exercise 2A

1. (i)

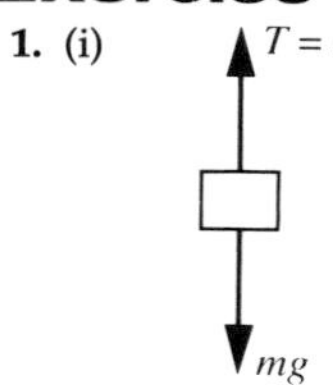

(ii) 6.125 N m^{-1} (iii) 12.25 N

2. (i) 20 N (ii) 20 N (iii) Tension required to double the length is the same
3. (i) 0.03 N (ii) 0.04 m (iii) 0.45 m (iv) 0.1125 N
4. (i) 4.9 N (ii) 61.25 N m^{-1} (iii) 0.625 kg (iv) 1.2 m
5. (i) 30 N (ii) 3 kg (iii) 300 Nm^{-1} (iv) String becomes fully compressed with fewer than 7 bricks

Exercise 2B

1. (i) 60 N (ii) 0.04 m (iii) 2400 N 2. (i) 12 000 N (ii) 80 000 Nm^{-1} (iii) 0.05 m (iv) 0.02 m
3. (i) 7000 N (ii) 7000 N, 10 000 Nm^{-1} (iii) 4000 N
4. (i) 197.77 N (ii) 653.78 N (iii) 118% (iv) 51.6 m
5. (i) $T = F, \frac{F}{k_1}, \frac{F}{k_2}$
6. (i) $0.3 - x$ (ii) $16x$, $25(0.3 - x)$ (iii) 0.183 m
7. (i) $2.2 - h$, $h - 1.2$ (ii) $44 - 20h$, $30h - 36$ (iii) 1.2 m (iv) 20 N, 0 N
8. (i) $\frac{l_0}{\lambda} mg\sin\alpha$ (ii) (a) $\frac{l_0}{\lambda} mg\,(\mu\cos\alpha + \sin\alpha)$ (b) $\frac{l_0}{\lambda} mg\,(\sin\alpha - \mu\cos\alpha)$
9. (i) 1000 N (ii) 0.024 m (iv) $R = mg - 5000(0.17 - h)$

10. (ii) 0.30625 m (iii) an elastic string is unlikely to pass smoothly over a peg

11. (i)

Exercise 2C

1. (i) 0.1 J (ii) 0.001 J (iii) 0.4 J (iv) 0 J
2. (i) 4 J (ii) 0.25 J (iii) 1 J (iv) 0.0625 J
3. (i) 0.75 J (ii) 5.48 ms^{-1}
4. (i) 0.00667 J (ii) 0.577 ms^{-1}
5. (i) 5×10^4 J (ii) 7.07 ms^{-1} (iii) 1.29 m (iv) 7.07×10^4 N, 35.4 ms^{-2}, (v) Truck moves back along the rail with same speed if other forces ignored
6. 0.433 m, 0.067 m

Exercise 2D

1. (i) $5h^2$ (ii) 0.2 gh (iii) 0.392 m
2. $2l_0$ m
3. $l_0\left(1 + \dfrac{mg}{\lambda}\right)$
4. (ii) 0.1294 m
5. 0.463 m
6. (i) 1 J (iii) 0.2 m
7. (i) 0.1098 m (ii) 0.149 m
9. (i) $\arccos\left(\dfrac{g}{w^2 l}\right)$ (ii) $\arccos\left(\dfrac{g(k - mw^2)}{w^2 l_0 k}\right)$

Exercise 3A

1. (i) 5 m (ii) 16 s (iii) 5, $\frac{\pi}{8}$ (iv) (a) 0 (b) $\frac{5\pi}{8}$ ms^{-1} (c) 0
2. (i) 2, $\frac{\pi}{6}$ (ii) 40 m (iii) $\frac{\pi}{3}$ ms^{-1} (iv) $\pm\sqrt{3}$ m
3. (i) 0.2 s, 0.08 m (ii) 0.08 m, 10π, 0, 0.8π ms^{-1} (iii) max speed 2.51 ms^{-1} (iv) 7.999 cm
4. (i) 2.5, 4 (ii) 40 $\text{cm}^{-2} \rightarrow$ centre (iii) 6 cms^{-1} (iv) (a) 6.89 cm (b) 12.53 cm
5. (i) $v = a\omega \cos\omega t$ (ii) velocity can be in either direction (iii) $\ddot{x} = -a\omega^2 \sin\omega t$
6. (i)

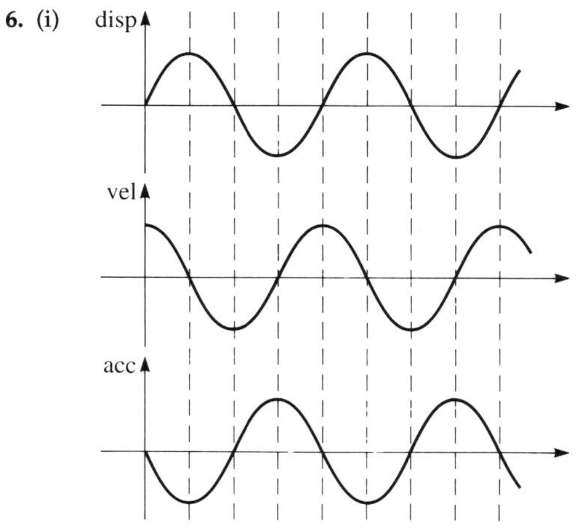

(ii)

Displ	Vel	Accel
max	0	min
0	max	0
0	min	0
min	0	max

7. (i)

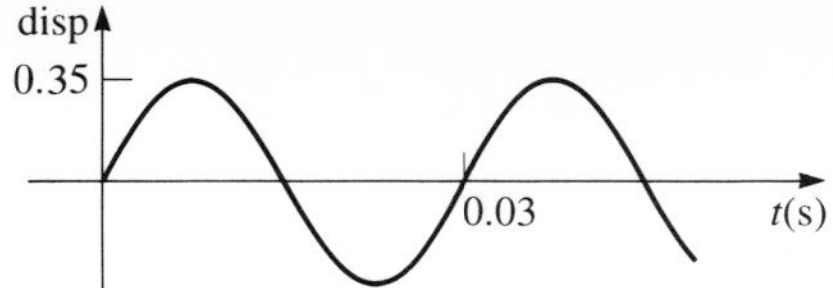

(ii) 33.3 Hz (iii) 73.3 ms^{-1}

8. (i) T = 0.0333 s (ii) 0.942 ms^{-1} (iii) 177.65 ms^{-2}

9. (i) 3.33 s (ii) 1.126 m (iii) 2.12 ms^{-1}

10. (i) 2.67 ms^{-1} (ii) 838.92 ms^{-2} (iii) 1.891 ms^{-1}

11. (i) 0.0005 s, 12566 (ii) 25.13 ms^{-1}, 3.16×10^5 ms^{-}
(ii) 22.12 ms^{-1}, 0.1728 ms^{-1}

12. (i)

3520 Hz	22116.8
1760 Hz	11058.4
880 Hz	5529.2
220 Hz	1382.3
110 Hz	691.15
55 Hz	345.6
27.5 Hz	172.8

Exercise 3B

1. (i)

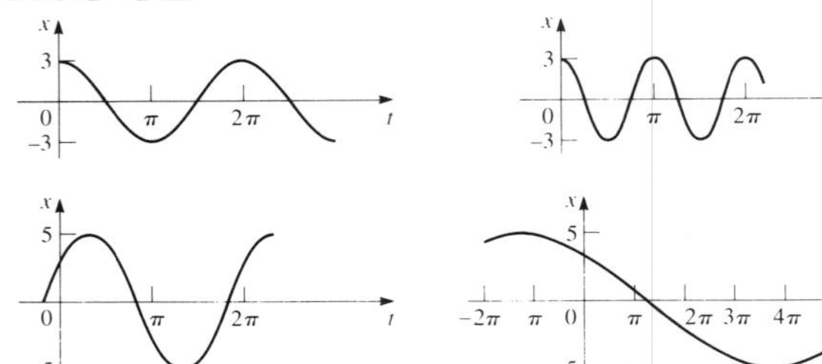

(ii) (a) $a = 3, T = 2\pi, f = \frac{1}{2\pi}$ (b) $a = 3, T = \pi, f = \frac{1}{\pi}$ (c) $a = 5, T = 2\pi, f = \frac{1}{2\pi}$
(d) $a = 5, T = 10\pi, f = \frac{1}{10\pi}$

2. (i) $x = 5\cos\left(\frac{2\pi t}{3}\right)$ (ii) $x = 6\sin\pi t$

(iii) $x = 2\cos\frac{2\pi t}{5} + 2\sqrt{3}\sin\frac{2\pi t}{5} = 4\cos\left(\frac{2\pi t}{5} - \frac{\pi}{3}\right)$

(iv) $x = 1.5\cos\frac{\pi t}{2} + \frac{12}{\pi}\sin\frac{\pi t}{2}$

3. (ii) amplitude = 60 mm   period = 1.05 s   frequency = 0.95 Hz

4. (i) $a = 5$, $\omega = \frac{\pi}{4}$, $\varepsilon = 0.7954$ (ii) $\delta = 0.7754$ (iii) b = 3.5355

5. (i) $y_0 = 8$ (ii) $a = 3.002$, $\omega = \frac{\pi}{6}$, $\varepsilon = \frac{\pi}{6}$

6. (i)

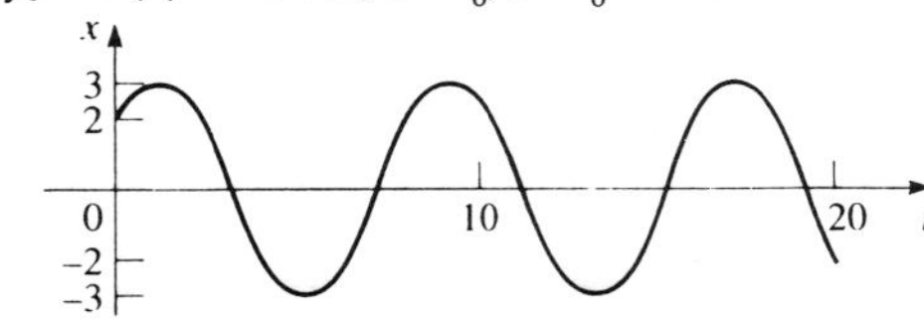

(ii) $a = 3$, $\omega_1 = \frac{\pi}{4}$, $\varepsilon = 0.73$ (iv) $b = 3$, $\omega_2 = \frac{\pi}{4}$, $\delta = \varepsilon - \frac{\pi}{2} = -0.84$

7. (i) (ii)

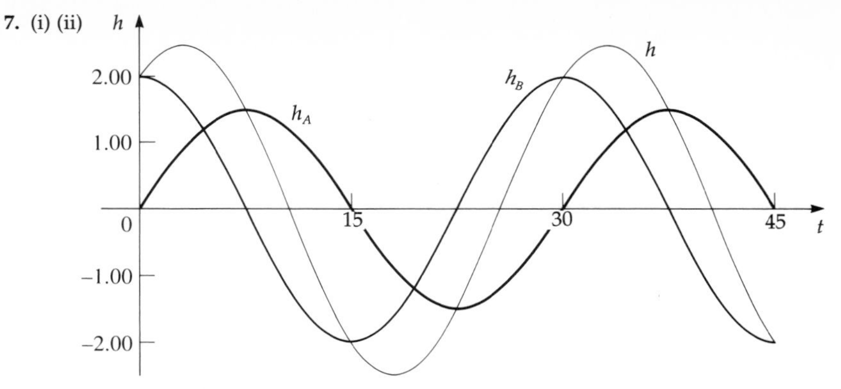

(iii) $a = 2.5$, $\omega = \frac{\pi}{15}$, $\varepsilon = 0.9273$

8. (i)

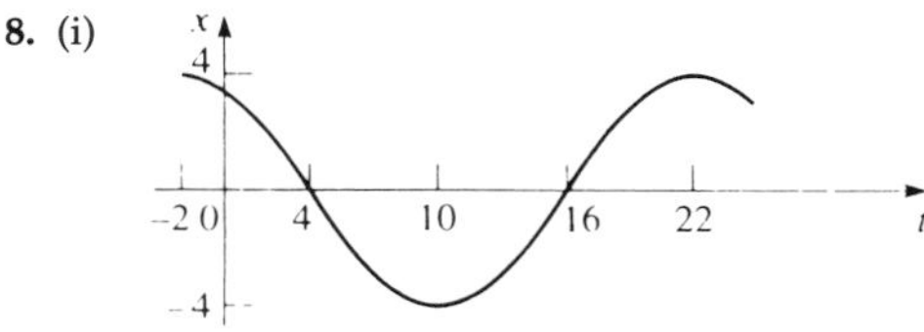

(ii) $a = 4$, $\omega = \frac{\pi}{12}$, $\varepsilon = \frac{2\pi}{3}$ (iii) $p = -2$, $q = 2\sqrt{3}$

9. (i) 85, 65, $\frac{2\pi}{11}$ (ii) 112, 42

10. (i) shm, mean h, amplitude a, period $\frac{2\pi}{\omega}$ (ii) 6, 4, $\frac{\pi}{6}$, $\frac{\pi}{6}$ (iii) 0.647 mhr^{-1}, 1.043 mhr^{-2} (iv) 2.094 mhr

11. (i) 0.0598 s, 0.5 (ii) $\beta = 1.5 + 1.5\sin(1.047t)$

Exercise 3C

1. (i) 0.0873 rad (ii) 3.48 s (iii) $a = 0.0873$, $\omega = 1.807$, $\varepsilon = 0$ (iv) 0.474 ms^{-1}
2. (i) 0.993 m (ii) (a) none, (b) multiplies by $\sqrt{2}$ (c) none (d) multiplies by $\sqrt{6}$
3. (i) $\ddot{\theta} = -\frac{g}{4}\theta$ stone is a particle, no air resistance, θ is small (ii) 4.01
4. (i) bouyancy = weight, $k = \frac{mg}{l}$ (ii) $F = \frac{mgx}{l}$, $\ddot{x} = -\frac{g}{l}(x - l)$
5. (ii) $x = 0.2$, $\dot{x} = 0$ when $t = 0$ (iv) $a = 0.2$, $f = 0.356$ (v) 0.447 ms^{-1}
6. (ii) $\ddot{x} = -600x$ (iii) $t = 0$, $x = -0.02$ (iv) $x = 0.02\cos(\sqrt{600}t + \pi)$ (v) 3.90 Hz (vi) the greatest speed = 0.49 ms^{-1}, the greatest acceleration = 12 ms^{-2}
7. (i) $\frac{151}{4}$ (ii) $\frac{\lambda}{4}$ (iii) $\ddot{x} = -\frac{4\lambda}{3lm}x$ (iv) $x = \mu\sqrt{\frac{3l}{4\lambda m}}\sin\left(\sqrt{\frac{4\lambda}{3lm}}t\right)$

   (v) $a = \mu\sqrt{\frac{3l}{4\lambda m}}$, $f = \sqrt{\frac{\lambda}{3lm}} \times \frac{1}{\pi}$
8. (i) 0.025 m (ii) $7.5 + 300x$ (iv) $x = 0.02\cos(20t)$ (v) 3 cm > 0.025 m so string is not in tension throughout
9. (i) $l_0 + \frac{mgl_0}{2\lambda}$ below the ceiling, $2l_0 + \frac{2mgl_0}{\lambda}$ below the ceiling.

   (ii) $\ddot{x} = \frac{-2\lambda x}{ml_0}$, $\ddot{x} = -\frac{\lambda x}{2ml_0}$ (iii) $\ddot{x} = \frac{-2\lambda x}{ml_0}$, $\pi\sqrt{\frac{2ml_0}{\lambda}}$, $2\pi\sqrt{\frac{2ml_0}{\lambda}}$
10. (i) (a) 2.1 ms^{-1} (b) 0.87 m (c) 0.022 N (ii) 6 s

Exercise 4A

1. (a) $\frac{104}{3}\pi$ (b) 8π (c) $\frac{652}{3}\pi$ (d) $\frac{2}{3}\pi$ (e) $\frac{9}{2}\pi(e^3 - 1)$ (f) $\frac{\pi^2}{4}$
2. (a) $\frac{96}{5}\pi$ (b) 18π (c) $\frac{28}{3}\pi$ (d) 24π

3. (i)

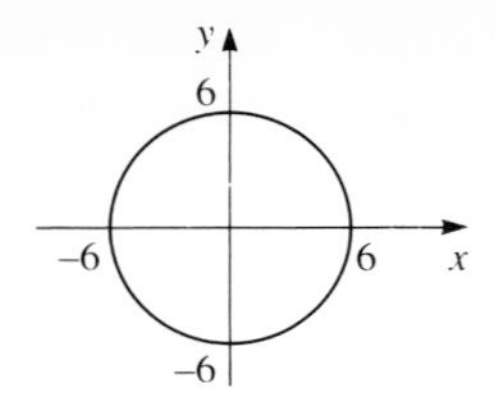

(ii) $0.235l$

4. (i)

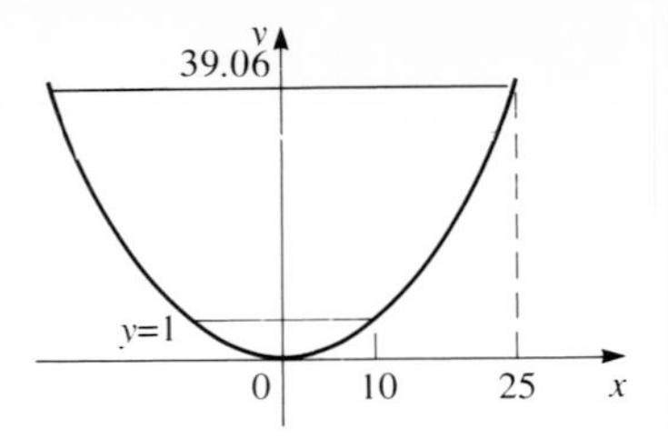

(ii) $45l$

5. (i) $y = \dfrac{rx}{h}$

6. (i)

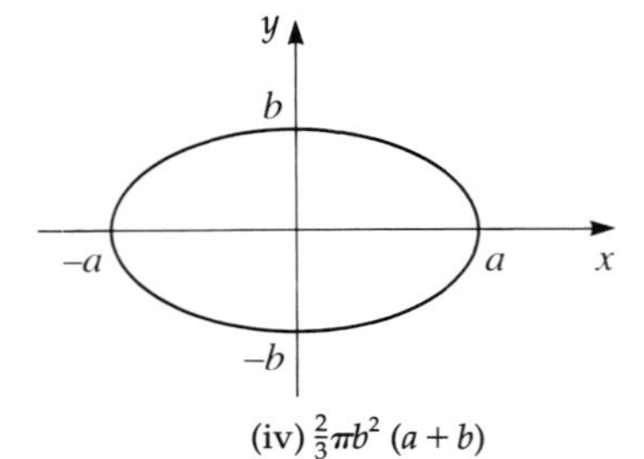

(ii)

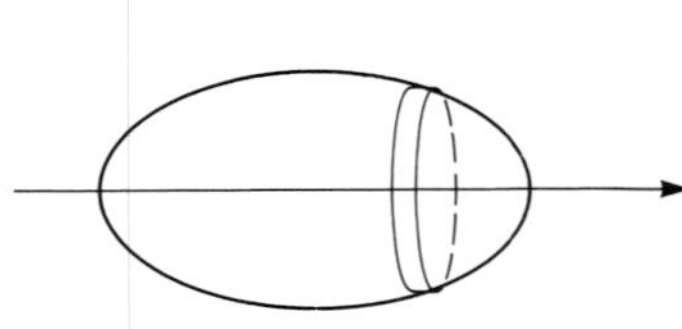

(iv) $\frac{2}{3}\pi b^2 (a + b)$

7. (i)

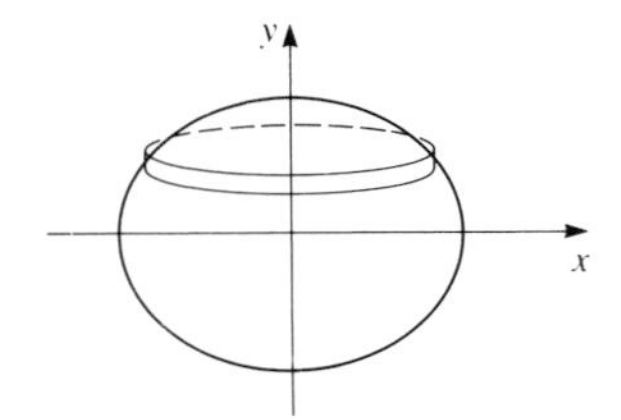

8. just over half **9.** (iii) 3.2×10^5 m^3

10. (i)

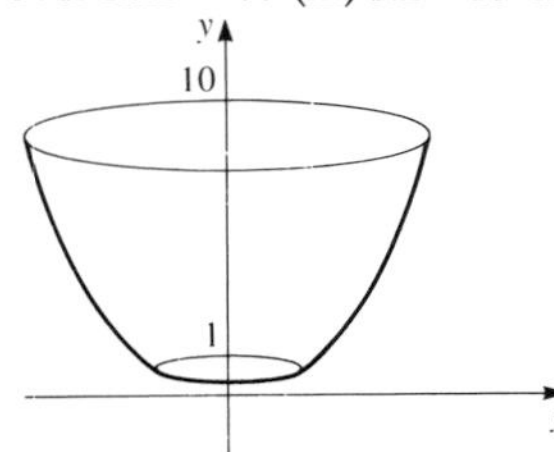

(ii) (a) 2.36 (b) 5.83 (c) 9.03

11. (i) A(−2, 0) B(2, 0) (ii) 11.29 (iii) 4765 cm^3

Exercise 4B

1. (i)

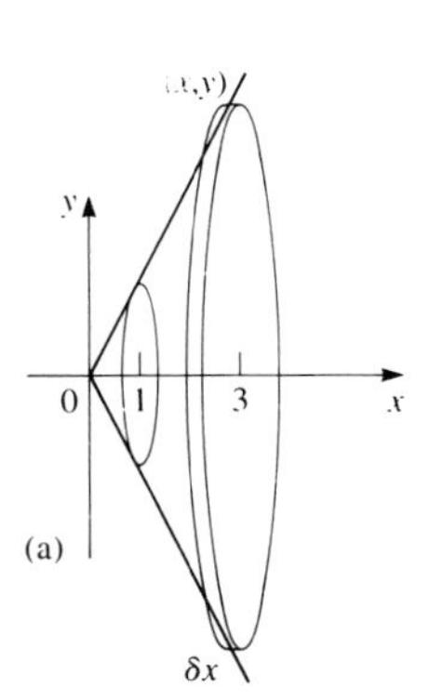

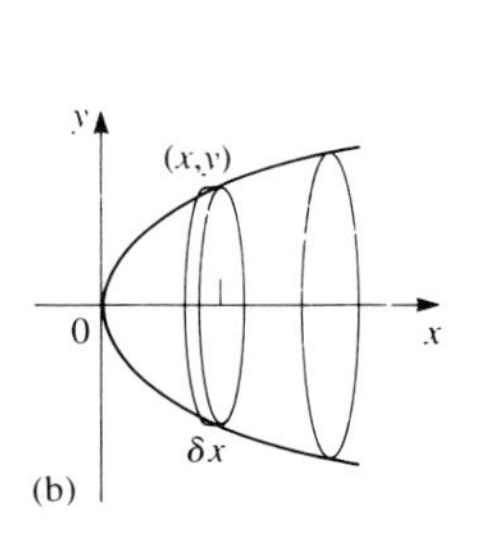

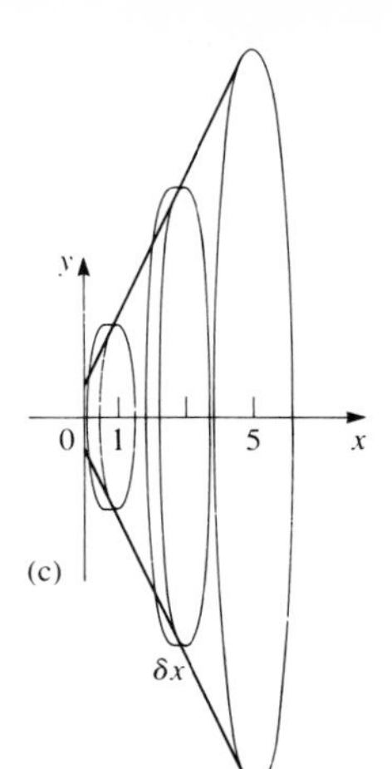

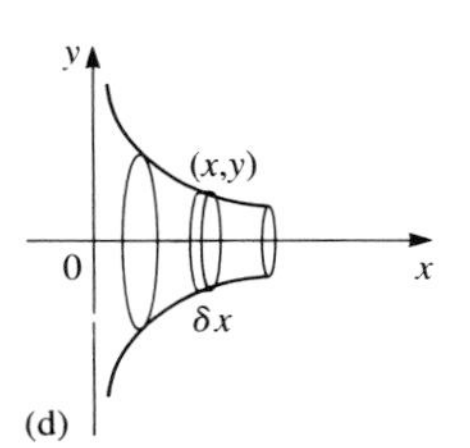

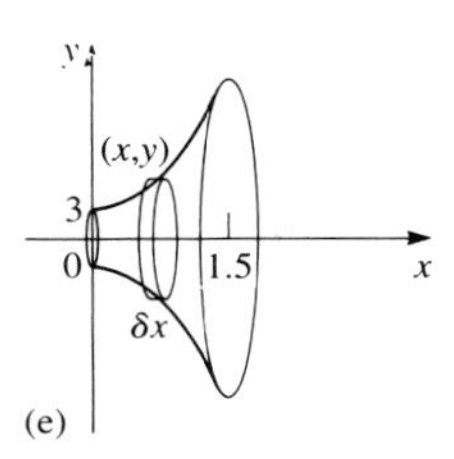

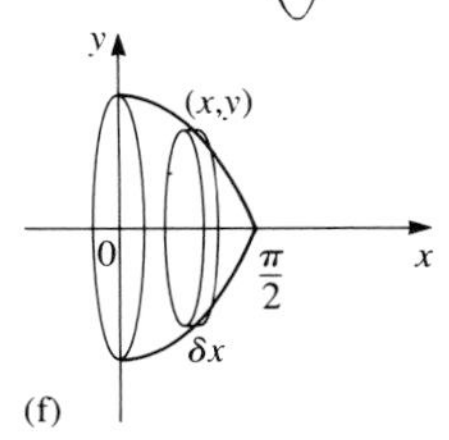

(ii) $\bar{y} = 0$, $\bar{x} =$ (a) 2.31 (b) 2.67 (c) 3.69 (d) 1.65 (e) 1.08 (f) 0.47

2. (i)

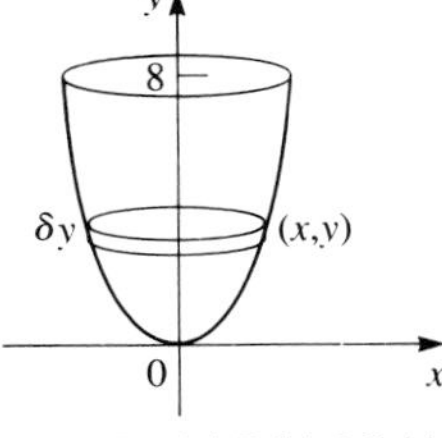

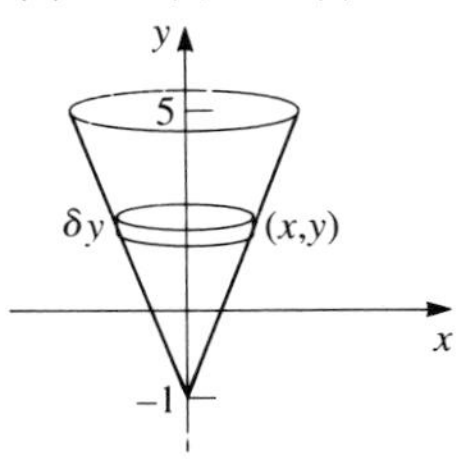

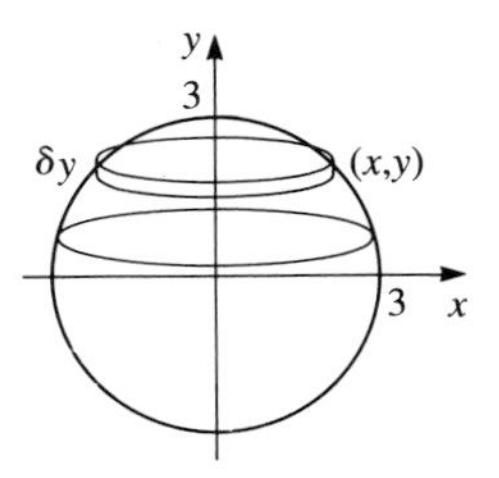

(ii) $\bar{x} = 0$, $\bar{y} =$ (a) 5 (b) 3.5 (c) 1.71

3. (i)

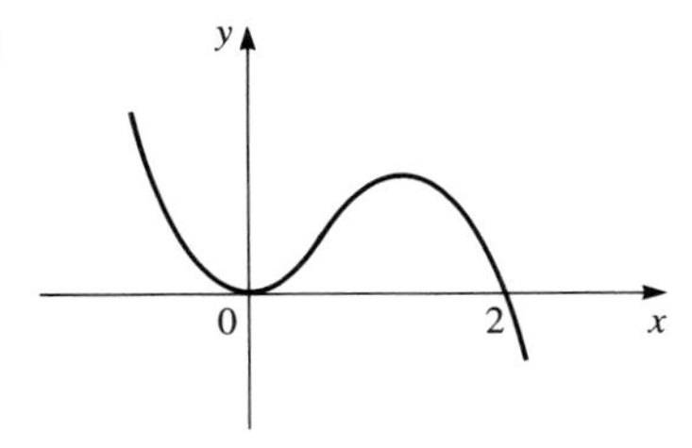

(ii) $\dfrac{128}{105}\pi$ (iii) 1.25

4. (i)

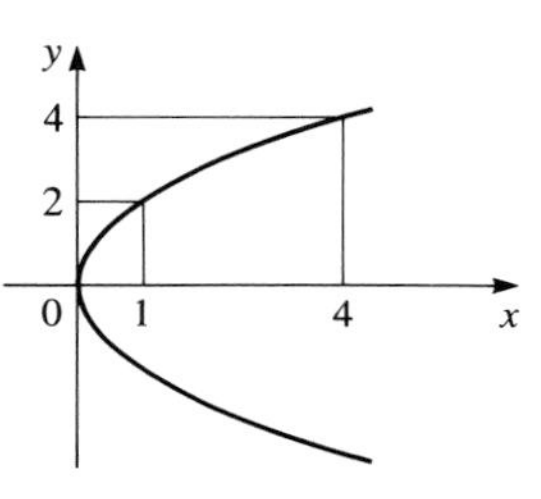

(ii) $v = 30\pi$, $\bar{x} = 2.8$ (iii) (a) 2, 4 (b) $\dfrac{62\pi}{5}$ (c) $\bar{y} = 3.39$

5. (i)

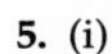

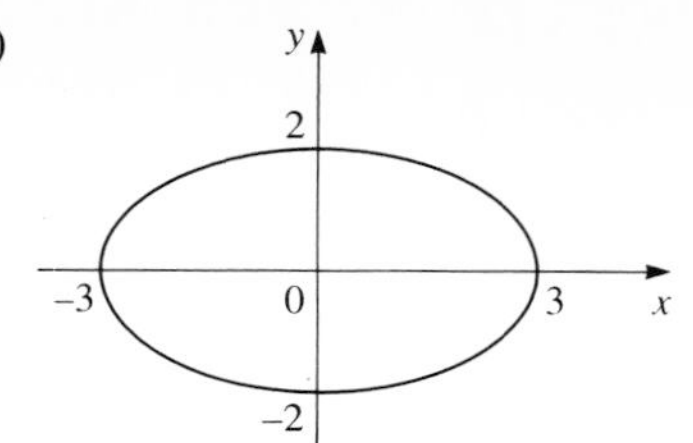

(ii) $V = 8\pi$, $\bar{x} = 1.125$

6. 3.9 cm

7. (i)

	Clown	Hemisphere	Cone
Mass	$\frac{\pi r^2}{3}\rho(2r+h)$	$\frac{2}{3}\pi r^3\rho$	$\frac{1}{3}\pi r^2 h\rho$
Dist. of c.o.m. above 0	$\bar{y}$	$-\frac{3}{8}r$	$\frac{1}{4}h$

(ii) $\frac{1}{3}h \times \frac{1}{4}h - \frac{2}{3}r \times \frac{3}{8}r = (\frac{2}{3}r + \frac{1}{3}h)\bar{y}$

(iii) $\sqrt{3}$ (iv) upright

8. (i)

	Container	Base	Cylinder walls	Cone
Mass	$132\pi\rho$	$9\pi\rho$	$108\pi\rho$	$15\pi\rho$
Dist. of c.o.m. above O	$\bar{y}$	0	9	$19\frac{1}{3}$

(ii) 9.56 cm (iii) 0.30 rads, 17.42° (iv) The centre of mass will be lower so it is less likely to topple

9. (i) $V = \dfrac{\pi h^2}{2k}$ (ii) $\frac{1}{3}h$ (iv) 7.3 cm (v) 0.39 rads, 22.4°

Exercise 4C

1. (i)

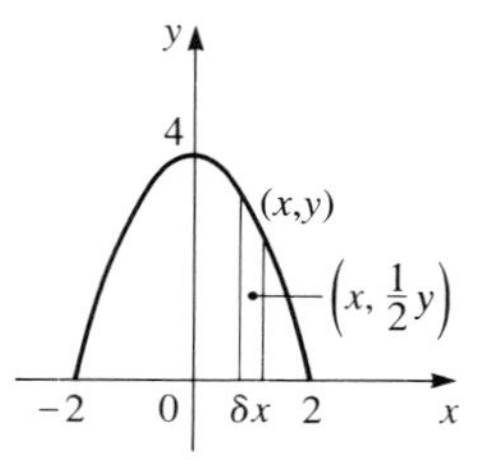

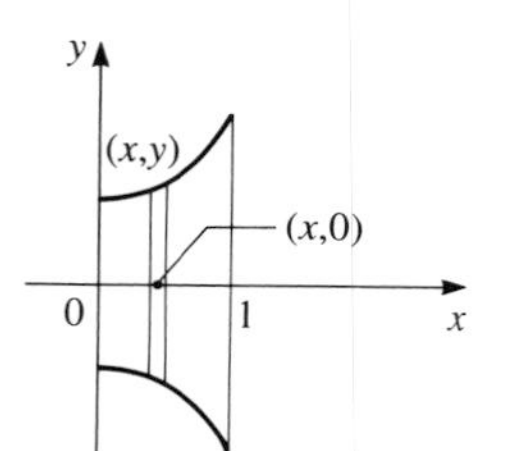

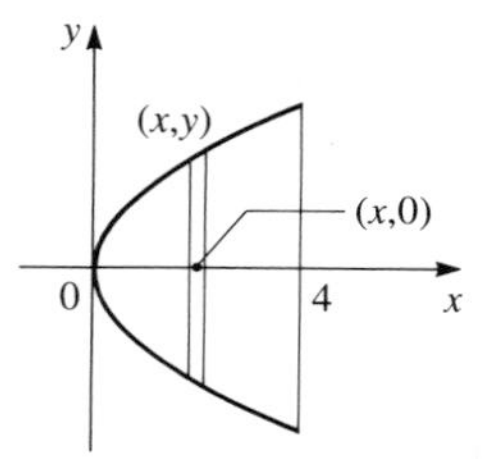

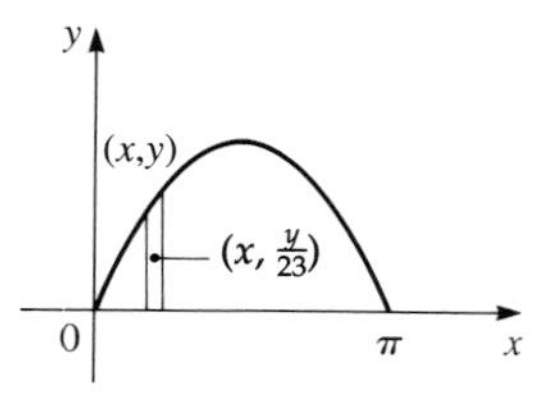

(iv) (a) $(0, 1.6)$ (b) $(0.58, 0)$ (c) $(2.4, 0)$ (d) $(\frac{\pi}{2}, \frac{\pi}{8})$

2. (i) 36 cm^2 (ii) $y = -2x + 12$ (iii) $(2, 4)$

3. (i)

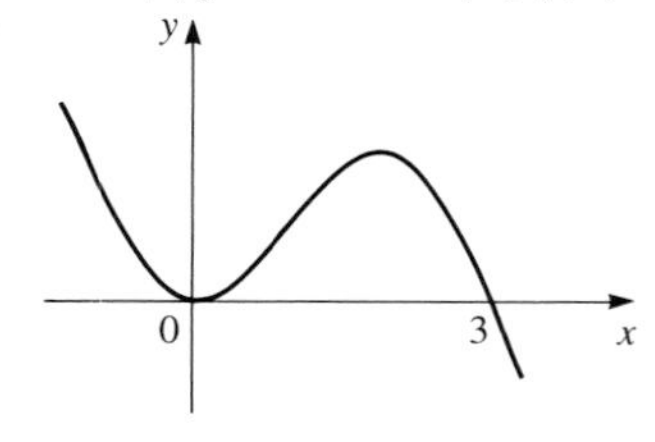

(ii) $(1.8, 1.54)$

4. (i) (ii) $4-x$, $(4-x)\delta y$, $\frac{4+x}{2}$ (iii) (2.4,0)

5. (i) outer semi circle $= 2\pi$, $(1+\frac{8}{3\pi}, 4)$ inner semi circle $= \frac{\pi}{2}$, $(1+\frac{4}{3\pi}, 4)$ rectangle $= 6$ (0.5, 3)
(ii) (1.16,3.44)

6. (i)

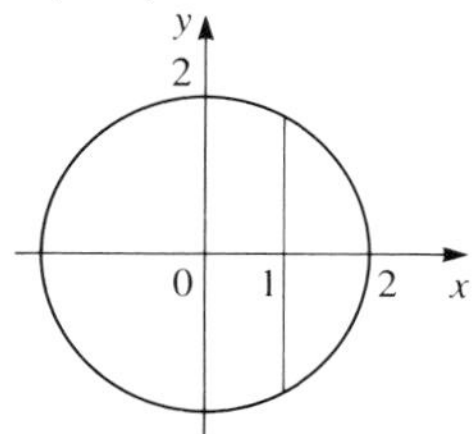

(iii) 1.41 (iv) 1.09 from centre (v) 32.2°

Exercise 5B

1. (a) All L^2T^{-2} consistent (b) All L consistent (c) All MLT^{-2} consistent (d) L.H.S. MLT^{-2} R.H.S. ML^2T^{-2} inconsistent (e) L.H.S. ML^2T^{-2} R.H.S. MLT^{-1} inconsistent (f) All MLT^{-2} consistent (g) L.H.S. ML^2T^{-2} R.H.S. MLT^{-2} inconsistent (h) All ML^2T^{-2} consistent (i) Two terms L^2T^{-2} but $[x^2] = L^2$ inconsistent (j) L.H.S. ML^2T^{-3} R.H.S. L^2T^{-3} inconsistent

2. (i) $[v] = LT^{-1}$, $[h] = L$, $[\rho] = ML^{-3}$, $[g] = LT^{-2}$ (ii) $\alpha = \frac{1}{2}$, $\beta = 0$, $\gamma = \frac{1}{2}$, $v = k\sqrt{hg}$ (iii) Deep water (iv) No difference

3. (i) $t = kd^{\alpha}g^{\beta}$ (ii) $t = k\sqrt{\frac{d}{g}}$ (iii) $k = 150$ (iv) 36.7 s

4. (i) ML^2T^{-2} (ii) $E = km^{\alpha}r^{\beta}\omega^{\gamma}$ (iii) $E = kmr^2\omega^2$ (iv) $E = \frac{1}{2}mr^2\omega^2$ (v) 5184 J

5. (i) e = velocity/velocity (ii) $\alpha = 2$, $\beta = 0$, $\gamma = -1$, $h = \frac{kv^2}{g}$ (iii) h is independent of m so both bounce to the same height (iv) because e is dimensionless and is included in k

(v) $h = \frac{e^2v^2}{2g}$ (vi) $k = \frac{e^2}{2}$ is dimensionless

6. (i) $M^{-1}L^3T^{-2}$ (ii) $t = k\sqrt{\frac{r_0^{\ 3}}{Gm}}$ (iii) Yes (iv) $t = \frac{k'}{\sqrt{G\rho_0}}$

7. (i) $x + y = 3$ (ii) $R = 6.72 \times 10^6 \times \frac{r^4}{a}$ (iii) 0.213 m^3s^{-1}

8. (i) $v = \frac{kmg}{r\eta}$ (ii) $m = \frac{4}{3}\pi r^3\rho$ (iii) $V = \frac{k_1r^2\rho g}{\eta}$ (iv) 0.113 cm

9. (i) π = Circumference of a circle ÷ its diameter, dimensionless (ii) $M^{-1}L^3T^{-2}$ (iii) 1.8×10^9 (iv) π (dimensionless) is independent of units, G (not dimensionless) is dependent on its units

10. (i) $f = \frac{k}{l}\sqrt{\frac{T}{m}}$ (ii) $f = \frac{1}{2l}\sqrt{\frac{T}{m}}$ (iii) 316.2 cycles s^{-1}

11. (iii) MT^{-1} (iv) Dimensionless (v) $\varepsilon = \frac{kcu}{mg}$

Exercise 5A

Quantity	Formula	Dimensions	SI Unit
Area	$l \times w$	L^2	m^2
Volume	$l \times w \times h$	L^3	m^3
Speed	$\frac{d}{t}$	LT^{-1}	ms^{-1}
Acceleration	$\frac{v}{t}$	LT^{-2}	ms^{-2}
g	acceleration	LT^{-2}	ms^{-2}
Force	$F = ma$	MLT^{-2}	N, newton
Weight	mg	MLT^{-2}	N, newton
Kinetic Energy	$\frac{1}{2}mv^2$	ML^2T^{-2}	J, joule
Gravitational potential energy	mgh	ML^2T^{-2}	J, joule
Work	Fs	ML^2T^{-2}	J, joule
Power	Fv	ML^2T^{-3}	W, watt
Impulse	Ft	MLT^{-1}	Ns
Momentum	mV	MLT^{-1}	Ns
Pressure	$\frac{\text{Force}}{\text{Area}}$	$ML^{-1}T^{-2}$	Nm^{-2}
Density	$\frac{M}{V}$	ML^{-3}	kgm^{-3}
Moment	Fd	ML^2T^{-2}	Nm
Angle	$s = r\theta$	Dimensionless	(radian)
Angular velocity	$\frac{\text{Angle}}{\text{time}}$	T^{-1}	$rads^{-1}$
Angular acceleration	$\frac{\text{Angular velocity}}{\text{time}}$	T^{-2}	$rads^{-2}$
Modulus of elasticity	$T = \frac{\lambda x}{l_0}$	MLT^{-2}	N
Stiffness	$T = kx$	MT^{-2}	Nm^{-1}
Gravitational constant, G	$F = \frac{Gm_1m_2}{d^2}$	$M^{-1}L^3T^{-2}$	Nm^2kg^{-2}
Period	Time interval	T	s
Frequency	$\frac{1}{\text{period}}$	T^{-1}	Hz, hertz
Coefficient of restitution	$\frac{V \text{ separation}}{V \text{ approach}}$	Dimensionless	–
Coefficient of friction	$F = \mu R$	Dimensionless	–